SpringerBriefs in Molecular Science

History of Chemistry

Series Editor

Seth C. Rasmussen, Fargo, USA

For further volumes:
http://www.springer.com/series/10127

Alan J. Rocke · Hermann Kopp

From the Molecular World

A Nineteenth-Century Science Fantasy

by Hermann Kopp

Translated, annotated and introduced by
Alan J. Rocke

 Springer

Alan J. Rocke
Department of History
Case Western Reserve University
Cleveland 44106
USA

Hermann Kopp (1817–1892)
Heidelberg
Germany

ISSN 2191-5407
ISBN 978-3-642-27415-2
DOI 10.1007/978-3-642-27416-9
Springer Heidelberg New York Dordrecht London

e-ISSN 2191-5415
e-ISBN 978-3-642-27416-9

Library of Congress Control Number: 2011945823

Printed on acid-free paper

Springer is part of Springer Science+Business Media (www.springer.com)

Preface

The German text for this English translation of Hermann Kopp's *Aus der Molecular-Welt* was provided by a copy in the Special Collections Department of Kelvin Smith Library on the campus of Case Western Reserve University, and I thank Norma S. Hanson for invaluable assistance, in this instance as so often in the past. That copy is of the third printing (1886) of this charming little book, but I have determined, by examining complete scanned copies of the other two editions in Google Books, that all three editions of this title were printed from a single set of plates, only the front matter distinguishing these printings one from another. In this translation edition, I substituted here the wording of the title page of the first (private) printing of March 1882 for that of the 1886 edition.

Further details of the interesting printing history of *Aus der Molecular-Welt* may be found in the introduction's footnotes no. 34, 36, 46, and 48 below. For most informative correspondence I am indebted to the kindness of Carl Winter of Heidelberg, great-grandson of Carl Winter (1836–1901), who owned Carl Winters Universitätsverlag Heidelberg, the family firm that published all three of these editions. I also thank Roland Adunka, director of the Auer von Welsbach Museum in Althofen, Austria, for helpful correspondence concerning Robert Bunsen's copy from the first printing of the book, held by the Welsbach Museum. My introduction is, in part, an expanded and revised version of a portion of Chap. 9 of *Image and Reality: Kekulé, Kopp, and the Scientific Imagination*, published by the University of Chicago Press in 2010. An early version of another part of my introduction was orally presented at a 2009 colloquium in honor of Christoph Meinel in Regensburg; I thank Professor Meinel, Christine Nawa, Thomas Steinhauser, and Carsten Reinhardt both for kind hospitality and for their helpful comments on that occasion.

Part of the research for this project was supported by Scholar's Award number SES-0618093 from the National Science Foundation, and part by the Henry Eldridge Bourne endowment, College of Arts and Sciences, CWRU. I am deeply grateful to German archivists and colleagues who provided access to manuscript sources, especially the Handschriftenabteilung of the Bayerische Staatsbibliothek (Munich); the archive of the Max-Planck-Gesellschaft (Berlin); the archive of the

Berlin-Brandenburgische Akademie der Wissenschaften (Berlin); the archive of the Deutsches Museum (Munich); the August-Kekulé-Sammlung of the Institut für Organische Chemie, Technische Universität Darmstadt (in 2010, this collection was wholly transferred to the Deutsches Museum Archive); the archive of Vieweg-Verlag (Braunschweig); the Universitätsarchiv Heidelberg; the Deutsches Literaturarchiv (Marbach/Neckar); the Bunsen-Bibliothek of the Auer von Welsbach Museum (Althofen); and the Royal Society of Chemistry (London). Kopp's German is occasionally challenging fully to comprehend, much less accurately to translate, and many of his references and allusions were initially more or less obscure; I am indebted to Ursula Klein, Susanne Vees-Gulani, and Christine Nawa for advice about the meaning of a few passages over which I was brought to the brink of despair. Last but by no means least, I most warmly thank Elizabeth Hawkins and Seth Rasmussen, whose professionalism made the last stages of this project a pleasure.

It is a perennial challenge for any translator to create new language that matches the original as closely as possible in both meaning and style; compromises between literal equivalence and graceful expression are inevitable. Suffice it to say that I have tried to steer my philological craft as wisely as I was able. Where I have perceived Kopp as being humorously obtuse or self-referentially prolix, I have not hesitated to produce English in a similar style. For all translations from manuscript sources cited in the introduction, the original German is provided in footnotes, reproduced *literatim et verbatim*. As was usual for German printed scientific works during the nineteenth century, eszett is here normally written out (ß = ss) when reproduced from printed sources. I use the following abbreviations for locations of manuscript sources:

ADM	Archiv des Deutschen Museums, Munich (with document number)
AKS	August-Kekulé-Sammlung, Institut für Organische Chemie, Technische Universität Darmstadt (located now in the Deutsches Museum Archive, Kekulé-Nachlaß)
BBAW-CB	Sammlung Chemiker-Briefe, Berlin-Brandenburgische Akademie der Wissenschaften, Berlin
BBAW-W	Sammlung Wachs, Berlin-Brandenburgische Akademie der Wissenschaften, Berlin
BSB	Sammlung Liebigiana IIB, Bayerische Staatsbibliothek, Munich
DLA	Sammlung Winter-Universitätsverlag Heidelberg, Deutsches Literaturarchiv, Marbach/Neckar
MPGA	Kopp file, II./38A, Nr. 4-44, Max Planck Gesellschaft Archiv, Berlin
RSC	Henry E. Roscoe Collection, Royal Society of Chemistry, London
UAH	Universitätsarchiv Heidelberg
VA	Kolbe file, Vieweg-Archiv, Vieweg Verlag, Braunschweig (with letter number)

Contents

Introduction: The Molecular World of Hermann Kopp

Hermann Kopp (1817–1892) is best remembered today as a historian, but during his lifetime he was highly regarded as a chemist. Late in his career he wrote a fantasy about personified molecules. Published in 1882, *Aus der Molecular-Welt* treated such subjects as atomic valence, molecular structure, the kinetic theory of gases, solution theory, and electrolysis, by imagining intimate details of what might actually be happening in the sub-microscopic world. The atoms and molecules we meet there have agency, personalities, sometimes even dialog. Filled with appealing tropes, humor, whimsical asides, and literary quotations, Kopp's short book provided an examination of the chemistry and physics of his day that was always light-hearted on the surface, but often surprisingly profound. There were others in Kopp's day who used a similar rhetorical strategy to write books intended mainly for children,[1] but this was a different sort of book, one directed to Kopp's professional colleagues. As we will see, his readers could readily perceive the important messages that lay beneath the fantasy and humor. Properly interpreted, the book constitutes a revealing tour of nineteenth-century debates concerning chemical theory, for Kopp's treatment here had an unbuttoned character that could not be found in more formal textbook accounts. As Friedrich Wöhler wrote to Kopp when he first received the book, "Your idea was superb, to clothe these serious chemical theories in humorous garb, to make them understandable."[2]

---

[1] Arabella Buckley's much-reprinted "fairy book" appears to have been the only one of this genre to predate the publication of Kopp's book [1, 2], but Buckley did not personify her fairies, and there is essentially no chemistry in it. Kopp's idea for his book was certainly independent of Buckley, for (as we will see below) he wrote an extended draft of his book in 1876, two years before Buckley's book appeared.

[2] "Es war ein vortrefflicher Gedanke von Dir, diese ernsten chemischen Theorien in spaßigem Gewand einzukleiden um sie dadurch verständlich zu machen." Wöhler to Kopp, 11 April 1882, MPGA.

A. J. Rocke and H. Kopp, *From the Molecular World*, SpringerBriefs in History of Chemistry, DOI: 10.1007/978-3-642-27416-9_1, © The Author(s) 2012

Aus der Molecular-Welt provides something of a master key to better understand not only Kopp's own personality and science, but also those of his predecessors and contemporaries. And it also suggests something important about the productive role of the imagination in the pursuit of science.

1 A Brief Biography

The son of a famous physician who had significant scientific expertise,[3] Hermann Kopp was born in Hanau, not far from Frankfurt, in the southernmost part of Electoral Hesse. After graduation from the classical Hanauer Gymnasium in 1835, Kopp studied for three semesters at the University of Heidelberg in the Grand Duchy of Baden, taking courses from the famous chemist Leopold Gmelin and the physicist Georg Wilhelm Muncke, as well as thoroughly mastering both Latin and Greek. He followed this with two more semesters of study at Electoral Hesse's own University of Marburg, where he earned his doctorate and *venia legendi* (teaching qualification) in the physical sciences in October 1838.[4] After a year at home in Hanau, he then traveled to Giessen, the university town of the Grand Duchy of Hesse, where he was welcomed into Justus Liebig's famous laboratory. He soon forged a firm friendship with the older Hessian chemist.

In March 1841 Kopp qualified as Privatdozent at Giessen, and began to teach theoretical chemistry, crystallography, and meteorology. Chemistry enrollments were rapidly expanding in Giessen at this time, largely as a consequence of Liebig's growing fame; but Kopp's teaching was also of the highest quality. In 1843 he was promoted to ausserordentlicher Professor, initially without salary.[5] Kopp's meager income dictated a spartan lifestyle; he lived in the home of the prominent physician Wilhelm Balser, probably a prior acquaintance of his father's.[6]

Kopp had a gentle persona, modest and unassuming; for instance, in his various histories he never once cited his own contributions to science. He also possessed a

---

[3] In addition to his renowned medical practice, J. Heinrich Kopp (1777–1858) was professor at the local Lyceum and co-founder of the Wetterauer Gesellschaft für die gesamte Naturkunde [3, 4].

[4] Before the arrival of Robert Bunsen in 1839, the University of Marburg had little scientific reputation compared with Heidelberg or Giessen. It is therefore likely that Kopp chose to transfer to his home-state University of Marburg for his doctoral examinations for purely logistical or pecuniary reasons, after his substantive education at Heidelberg.

[5] Generally, only full professors received salaries in nineteenth-century German universities; others in the teaching staff relied on student fees paid directly to them. The Grand Duchy of Hesse granted Kopp a small salary from 1846, engineered by Liebig to persuade Kopp not to consider a possible offer from the University of Kiel. Liebig already regarded the 28-year-old Kopp as "indispensable" [6, 1:269–73].

[6] Kopp lived (and worked in a small improvised laboratory) in the Balser family home until 1859, even after Wilhelm Balser's death early in 1846 and his own marriage in 1852.

quick wit as well as an astonishingly broad command of German and classical literature, and was widely admired for his entertaining conversation. He was a true old-school scholar, focused on the life of the mind, and he rarely concerned himself with his own material betterment. Considering all this, it is not surprising that he spurned an easier road by resolving to teach and to write on a passionate interest of his, the history of chemistry, though it was surely clear from the start that this was not the best strategy for career advancement. The first volume of his *Geschichte der Chemie* appeared in 1843, the fourth and last volume in 1847, shortly before his thirtieth birthday. There existed a few other histories of chemistry before Kopp's, but the subject had never been treated in such detail or with such careful primary-source scholarship. Kopp's fame as a historian was established. For decades thereafter he worked assiduously toward a new edition of this great treatise, but although in later years he was to publish three more heavy volumes on focused aspects of the history of chemistry, he never finished the task of revising and updating his general history.

In addition to his massive historical research, Kopp became prominent also by his important editorial and journalistic work. The great chemical periodical series of the 1820s, 1830s, and 1840s had been Berzelius's *Jahresberichte*, book-length annual critical reports on the contemporary chemical literature. This series was continued after Berzelius's death in 1848 by an editorial collective in Giessen, nominally headed by Liebig and Kopp [5]. Through the first fourteen annual editions of the new series, Kopp always did the lion's share of this tremendous task. His work in this arena was so thorough, professional, and impartial, that Liebig then decided to hand over to Kopp the chief editorial responsibility for his journal, the *Annalen der Chemie und Pharmacie*. In contemplating this step, Liebig as usual consulted his best friend, Wöhler, noting in favor of the move that Kopp's "talent, his orderly mind, his good judgement, and the breadth of his knowledge are unique." Wöhler agreed. "There could be no better editor than he. I have enormous respect for his quite exceptional mind. I only regret that he will in the end come to grief with all the correspondence and printer's ink" [6, 1:361]. From 1851 until 1871, "Liebig's" *Annalen* was actually run by Kopp.

When Liebig was called to the University of Munich in 1852, Kopp and his colleague Heinrich Will agreed to share Liebig's successorship as ordentlicher Professor in Giessen. This was not a recipe for an amicable collegial relationship. A year later Kopp voluntarily chose the title professor of theoretical chemistry, leaving the more general and lucrative professorship of chemistry to Will. The rise in income associated with Kopp's promotion, modest though it was,[7] enabled him to marry. His wife, Johanna née Tiedemann, was a niece of Friedrich Tiedemann, the great physiologist at Heidelberg. She was described as high-minded

---

[7] Kopp's salary as Ordinarius at Giessen was 1200 florins, or about 700 thalers per year [7, 414n., 8, 115], quite modest for Germany at this time, especially considering that his income from student fees would have been relatively low—Will having garnered the enrollments in the large general courses.

(hochsinnig) and sociable, and thoroughly devoted to her husband, as he was to her.[8] The couple had one daughter, Therese, born in 1856 [9, 413].

In Giessen, Kopp taught many of those who would become leaders of the next generation of chemists, including A. W. Hofmann, A. W. Williamson, B. C. Brodie, Adolf Strecker, August Kekulé, Edward Frankland, Emil Erlenmeyer, and Jacob Volhard. Marburg is only about 30 km from Giessen, so he was able to visit regularly there with Robert Bunsen and then (from 1851) with Hermann Kolbe, both of whom became close friends. With his extended collegial circle of friends, his total mastery of the literature and history of the field, his important editing duties, and his exceptional memory, judgement, and work ethic, there was no more central chemical personality in the middle decades of the nineteenth century than Hermann Kopp. He also published fundamentally important chemical research, which we will discuss in the next section.

Considering all this, it is not surprising that other German universities began to covet this extraordinary scholar. Between 1862 and 1870 Kopp received four offers from other universities, and at least two additional offers were nearly made—these calls and near-calls coming from many of the greatest German universities. It started with a possible call in 1862 to an unnamed university, probably Bonn, and apparently engineered by Liebig.[9] In 1863 there came an offer from Heidelberg, which Kopp accepted (more on this below). Thereafter, Hofmann repeatedly tried to persuade Kopp to join him in Berlin; official offers from the Prussian ministry came to Kopp in 1864 and again in 1865. The first of these calls failed only because the ministry's private advance assurance that Kopp would be given either a rent-free residence (Dienstwohnung) or at least a housing subsidy was not repeated in the official offer. A third official call to Berlin apparently would have been made in 1868 had Kopp shown any interest.[10] Finally, in 1870 Leipzig called Kopp to replace Otto Erdmann, but Kopp declined.[11]

The one offer that Kopp did accept was from Heidelberg. The call came because in 1863 Hofmann had actually first tried to get Bunsen to come from Heidelberg to Berlin, but Bunsen had declined, for he did not want to work under what he considered to be the shamefully ("schandbar") illiberal Prussian regime under Bismarck. Bunsen had been savvy; his condition to remain faithful to Baden was that the authorities allow him to bring a colleague to join him in Heidelberg as a second Ordinarius for chemistry in the Philosophical Faculty. He named Kopp.[12] Kopp was delighted to come to Heidelberg, partly because of his already close relationship with Bunsen. They had similar (physical–chemical) scientific interests, they had often visited with each other when Bunsen had taught at nearby

---

[8] On his fiftieth birthday Kopp wrote Liebig: "Gott lasse es meiner Frau gut gehen, dann wird sich Alles machen und tragen lassen"; letter of 31 October 1867, BSB.

[9] Kopp reminisced about this event in his letter to Liebig of 18 April 1864, BSB.

[10] Kopp to Liebig, 1 July 1864, BSB; also Hofmann [10, 511].

[11] Kopp to Kolbe, 23 and 28 June 1870, ADM 3521 and 3522.

[12] Bunsen to Roscoe, undated but clearly from the year 1863, RSC; reproduced in Stock [11, xxi n.].

Marburg during the 1840s, and Kopp had had repeated occasion to visit Bunsen at Heidelberg during the 1850s. Kopp arrived there in March 1864.

After the deaths of both his and his wife's fathers in 1858 and of his stepmother two years later, Kopp felt financial pressure, for in addition to concerns about his young daughter and his sometimes unhealthy wife, he now had become head of his birth family and trustee for the children of a deceased older sister [7, 397–98n.].[13] His move to Heidelberg, certainly a favorable move regarding the reputation of the university, did not advance his income, for he accepted the same rather poor salary there which he had been drawing in Giessen.[14] He would have tripled his salary had he accepted the subsequent calls to Berlin or to Leipzig, but there were non-monetary reasons why he really preferred Heidelberg [7, 406–9].[15]

If Kopp did not initially gain in pecuniary terms in Heidelberg, he was certainly appropriately honored. In 1867 he was named Geheimrat, and the next year he was made an Ehrenmitglied of the Deutsche Chemische Gesellschaft upon its founding, along with Liebig, Wöhler, Bunsen, and Kolbe. Although he was apparently never promised a proper laboratory, he also acquired better working conditions in Heidelberg. For the first few years he occupied a small structure in the garden of the medieval monastery that decades earlier had been converted into the university's chemical laboratory; in 1875 he was given laboratory space in the newly-built Friedrichsbau, which replaced the monastery [12, 23]. "I assure you," he wrote Liebig soon after his arrival, referring with distaste to the situation he had just left, "it means something when one is no longer young, to do such [scientific] work in a small chock-full room (the late Balser's bedroom), where one barely has room to stand, and where one must be one's own assistant, Aubel, and Aubel's own maid, even one's own mechanic."[16]

The need for additional income drove Kopp to furious activity. For years he spent the day in classroom and laboratory, and then much of the night writing [6, 2:336, 9, 408]. His editorial work and his voluminous historical and occasional

---

[13] Kopp to Liebig, BSB, 2 April 1860: "Eine Last von Pflichten ist jetzt auf mich …".

[14] Kopp to Badisches Kultusministerium, 23 November 1863, UAH, PA 1868, pp. 1, 3. Kopp recognized that his call to Heidelberg was unusual, for there was no actual vacancy at the time, and this may have inhibited him from demanding a market-level salary: Kopp to Hofmann, 10 March 1864, BBAW-CB. By comparison, Bunsen's salary was 2700 guilders, more than three times Kopp's, plus he was given a Dienstwohnung gratis [11, 536]. Hofmann came to Berlin at 2500 thalers; Liebig was making 5000 thalers in Munich.

[15] "Bunsen alone," he told Hofmann [10, 511–12], "holds me fast in Heidelberg." He, like his friend Bunsen, may not have relished moving to fast-paced, expensive, and politically less liberal Berlin; and he may have hesitated to trade Bunsen as close colleague with Kolbe at Leipzig.

[16] "… ich versichere Sie, es heißt etwas, wenn man nicht mehr jung ist, solche Arbeiten in einem vollgepfropften kleinen Raum (des sel. Balser's Schlafkammer) zu machen, wo man nicht mehr als eben den Platz zum Stehen hat, und wenn man sein eigener Assistant, Aubel und die dem Aubel beigegebene Putzfrau, nebenbei auch sein eigener Mechanicus sein muss" (Kopp to Liebig, 20 April 1864, BSB). Kopp was referring to Dr. Balser's unheated spare bedroom in Giessen, where he had fitted up a makeshift laboratory. Aubel was the factotum in the Giessen lab.

writings were pursued out of his own interests, certainly, but also for the royalties. He neglected his experimental research due to the press of writing, and also because of the expense of apparatus and supplies (for the state of Baden provided him no budget for his own research). His extensive correspondence contains laments, especially after about 1872, that he had not accepted this or that call, and regret that he had chosen to relinquish the editorship of the *Annalen* with its associated income.[17] Kopp was only too well aware that he had not played the self-interested academic game very well. "You know ... that when it rains pudding, I never seem to have a spoon," he commented ruefully to Liebig.[18] To Wöhler he lamented, "In my youth, mothers used to tell their daughters, 'Spin, girls, spin; your suitor sits therein,' in order to keep them working hard at the distaff. I spin until my fingers bleed; may Heaven grant that *my* suitor comes soon."[19]

Allowance must be made here for rhetorical coloration. Kopp eventually did manage to raise his remuneration to the level of Bunsen's (probably as a consequence of declining the call to Leipzig), even if no free housing came with the deal.[20] When Wöhler commented that this wasn't really a bad salary, Kopp pointed out that the salaries attached to recent calls of eminent new professors were half again as much; moreover, Heidelberg had become one of the most expensive cities in Germany and "the tavern of Europe."[21] But he could not deny that he had a very nice residence, a devoted wife and daughter, and a pleasant and productive working environment. He and his wife and daughter also enjoyed occasional long vacations with friends—most often Bunsen—especially to Italy. For all his amiable selflessness and good humor, Kopp seemed to enjoy grumbling to his friends. "You're quite right that I don't live in a sensible way," he wrote Wöhler. "But, que voulez-vous, que je fasse?"[22]

Meanwhile, colleagues noticed a change in his writing style. Early in his career he had written in clear, direct sentences that were capable of addressing large subjects efficiently; now he was hyper-self-critical and hyper-precise, writing in

---

[17] Kopp received 400 guilders per year in royalties for his editorship of the *Annalen* [7, 434]; this increased his annual salary after 1852 by a third. He told Wöhler (11 March 1873, BBAW-W) that he had relinquished the editing work in 1871 in the belief that he could replace that income, but that had not been possible, and he now had even higher expenses than expected.

[18] "Sie wissen ... daß mir beim Breiregnen gewöhnlich der Löffel fehlt." Kopp to Liebig, 10 October 1863, BSB.

[19] "'Spinn, Mädchen, spinn; der Freier der sitzt drin' sagten in meiner Jugend die Mütter um die Töchter fleißig am Rocken zu erhalten. Ich spinne, daß mir die Finger bluten; gebe der Himmel, ein Freier komme bald." Kopp to Wöhler, 15 March 1873, BBAW-W.

[20] On 27 March 1873 Kopp told Wöhler that his current salary was 2700 guilders per year (BBAW-W). It is likely that his decision to say no to Leipzig in 1870 had been the occasion for his raise in pay, which in turn enabled his decision to resign the editorship of the *Annalen* the following year. In more than doubling his income at Heidelberg, Kopp could figure he could bear the loss of 400 guilders annual royalty.

[21] Kopp to Wöhler, 27 March and 10 April 1873, BBAW-W.

[22] "Darin, daß ich nicht vernünftig lebe, hast Du sehr recht. Aber, que voulez-vous, que je fasse? [what do you want me to do?]" Kopp to Wöhler, 11 July 1873, BBAW-W.

enormous convoluted sentences that even well-practiced German colleagues often had difficulty untangling.[23] Both of these styles were perfectly correct in form, but the latter could be very hard to read. Kopp himself commented privately, in frustration rather than pride, on this change in his writing, and on his increasingly fussy (kleinlich) tendency that led to a kind of scholarly paralysis.[24] His students, especially those who did not have the advantage of native command of German, also had to pay very close attention to their professor's lectures. When he was a student of Kopp's, the philosopher Emile Meyerson for amusement would take out a stopwatch in lectures; the record (as he told the story to George Sarton) was a sentence that took Kopp no less than nine minutes to finish [10, 416–17].

Such concerns did not seem to diminish Kopp's scholarly output, nor his continued passionate engagement with teaching, nor the high regard of colleagues, students, and a wide readership. His *Beiträge zur Geschichte der Chemie* came out in two parts in 1869 and 1875; his *Entwickelung der Chemie in der neueren Zeit* appeared in 1873; and his *Alchemie in älterer und neuerer Zeit* was published in 1886, a total of over 2000 print pages just in these three titles. These works are still well worth consulting for those who concern themselves with the history of chemistry, as is his youthful 4-volume *Geschichte der Chemie*.

Kopp often commented in letters to friends on the pleasure and profit he derived from work in the classroom. He regarded teaching as a vital adjunct to research, for he thought that his lectures helped him to grasp difficult theoretical issues more fully; and he also regarded the historical approach as pedagogically useful, often even essential, for his students. This was why he so often touted the heuristic and epistemological importance of his historical pursuits—especially *recent* history of chemistry, which hardly any of his predecessors had even attempted.[25] Many colleagues agreed. When the publisher Alexander Macmillan declined to publish an English translation of Kopp's *Beiträge zur Geschichte der Chemie*, Henry Roscoe was shocked to hear Macmillan opine that "the British public does not care about history of chemistry!" This was a great disappointment, Roscoe reported to Kopp, "because I am sure that your 'Beiträge' are of the greatest importance for men of science."[26]

Until his last illness Kopp continued his heavy load of teaching and writing. His greatest pleasures were in work, in family, and in his good friends, of whom Bunsen was closest. The two men often took an evening walk together, through the

---

[23] Virtually all obituarists and biographers commented on the difficulty of Kopp's mature writing style. Hofmann wrote [10, 520]: "Indeed, Kopp's curious characteristic style prompted many complaints. Even his friends couldn't help occasionally teasing him in protest against his sentence constructions."

[24] Kopp to Liebig, 14 November 1865, 30 December 1871, and 5 February 1872, BSB.

[25] See, for example, Kopp to Liebig, 8 December 1872, BSB; Kopp to Hofmann, 25 January 1873, BBAW-CB; and Kopp to Liebig, 1 March 1865, BSB: "ich habe diesen Winter ordentlich gelehrt, so kann ich auch sagen, daß ich recht viel gelernt habe."

[26] Roscoe to Kopp, 18 January 1871, MPGA (the exclamation point was Roscoe's, relating, in English, his conversation with the publisher).

hills and parks of Heidelberg. Roscoe, who was personally close to both men, wrote that it was a comical sight to see the short and stout Kopp toddling along (einhertrippeln) beside the confident stride of the stately figure of Bunsen, Kopp's senior by six years and overtopping him in height by more than a foot. Kopp was "one of the most modest of men," Roscoe added, "but he had a sparkling humor, informed by the deepest understanding." Bunsen had a way with words, too. Roscoe reported that he once responded, tongue in cheek, to an inquiry concerning his old friend [13, 90–91]: "Yes, a small chemist by the name of Kopp did indeed live here. I knew him for thirty years, but I never understood a word he said." (Bunsen spoke the perfect Hochdeutsch of Hanover; Kopp spoke a Mittelhessisch dialelct.)

Toward the end Kopp struggled with poor health, resulting from both respiratory and kidney disorders. His illnesses led to his retirement on 1 July 1890, and he finally expired on 20 February 1892, after months of torment. Kopp's student and biographer T. E. Thorpe summed up a good life, lived fully [14, 776]: "To know Kopp, was to love him, and to love him was ... a liberal education."

2 Scientific Work

In nineteenth-century terms, Kopp was of a middle generation of chemists: younger by a few years than the great founding generation of Liebig, Wöhler, and Jean-Baptiste Dumas, and older by a few years than another group of radical innovators, Williamson, Frankland, Kekulé, and Dmitrii Mendeleev. Some of Kopp's close contemporaries, such as Bunsen, Hofmann, and Victor Regnault, were notably indifferent to theory; other members of his generation, such as Kolbe, Auguste Laurent, Charles Gerhardt, and Adolphe Wurtz, were just as powerfully attracted to theoretical ideas. And by "theory," I refer in particular to two broad intersecting areas of intense interest to many chemists during the middle decades of the century: chemical atomism, and investigations into the constitutions of molecules.

In this sense Kopp occupies a curious position. Judging solely from his published articles it would seem that he was anti-theoretical, for in his research he never clearly committed himself to one or the other side in these lively debates. His research was well adapted, as if intentionally, to avoid the hot topics, for Kopp's field was the then-marginal specialty of physical chemistry. Indeed, reading his scientific papers, overstuffed with mind-numbing tabulations of physical measurements, risks extreme boredom. When we get to the punch line of the article, all we usually find is an attempt to encapsulate the data in arithmetic rules that often lacked both precision and full generality. In these papers Kopp only rarely attempted to assess the larger physical or theoretical meaning of these formulae.

But this impression of antipathy to theory is misleading. Throughout his career Kopp's teaching specialty was (by his ardent preference) theoretical chemistry, and he was regarded by his German contemporaries as perhaps the greatest master of that field. When an eminent chemical author or a publisher needed to produce a section on theoretical chemistry for a larger work, Kopp usually received the call. Such was the case, for example, when Eduard Vieweg, the publisher of the monumental series entitled *Graham-Otto's ausführliches Lehrbuch der Chemie*, wanted a separate volume on physical and theoretical chemistry: Vieweg turned to Kopp (along with two coauthors) in 1857 [15], and as sole author Kopp wrote a second edition of the theoretical chemistry textbook in 1863 [16]. Again, when in 1868 Wöhler felt overmatched by the task, he asked his younger friend to write two major sections on theoretical chemistry for the fourteenth edition of Wöhler's own *Grundriss der unorganischen Chemie* [17, 1–50, 325–57]. Kopp fulfilled these requests, but when Hofmann asked him some years later to assist in the preparation of a new edition of his *Einleitung in die moderne Chemie*, Kopp was too far behind in his own work, and begged off.[27]

Kopp's deep engagement with theory motivated much of his scientific research, but in a way that was not immediately apparent. Kopp studied the relationship between physical properties and chemical compositions of substances [14, 18, 19, 2:17–19, 295–97]. He determined crystal forms, boiling points, densities, specific heats, thermal expansions, changes of volume consequent to changes of state, and solubility relations. He began these monumental labours at the age of 21, in the year he spent at home in Hanau between the awarding of his doctoral degree and his entry into Liebig's laboratory at Giessen [20]. The work occupied him for about forty years, until increasing age, financial stress, and the pressures of literary labour put an end to his experimental studies. His experiments were done using extremely simple apparatus, mostly in Giessen in a tiny improvised space. The accuracy of measurement he was able to achieve under these conditions was astonishing, as was the sheer number of determinations of physical properties of a galaxy of different substances.

Kopp was particularly focused on a property he called the "specific volume," which he defined as the atomic weight of an element (or the molecular weight of a compound), divided by the density of the substance. This property formed a close analogy to the so-called "atomic heat" of an element, defined by Petit and Dulong in 1819 as the element's atomic weight multiplied by its specific heat. Accordingly, Kopp's specific volumes were often called "molecular volumes"; today they are known as molar volumes. By comparing the precisely determined molecular volumes of enough carefully-chosen compounds, Kopp was to identify many interesting relationships and regularities, or near-regularities. His proximate goal was to construct, as he put it, a "stoichiometry of physical properties of chemical compounds," and he succeeded in going some good distance toward this goal. By 1863 he had included 150 compounds in his system.[28]

---

[27] Hofmann's request and Kopp's rebuff is in Kopp to Hofmann, 13 July 1882, BBAW-CB.

[28] Kopp to Liebig, 20 November 1854, 29 January 1855, 23 May 1863, and 10 October 1863, BSB.

Kopp's major results, announced in the mid-1850s, included the general comparative statement that a common difference in molecular constitution results in a proportional difference in molecular volume, as well as several generalizations relating to more specific aspects of chemical composition [21–23]. For instance, Kopp's data indicated that oxygen atoms that play distinguishable compositional roles in organic compounds—such as oxygen atoms in carbonyl groups versus oxygen atoms in hydroxyl groups—have distinguishable influence on molecular volume. By such means one could use Kopp's data to assist in the determination of molecular structures.

Kopp's work on boiling points, summarized in the same series of publications, had similar interest. He showed, for instance, that organic acids generally boiled about 40° C. higher than the corresponding alcohol and 82° higher than the isomeric ester; that functionally similar isomers such as ethyl formate and methyl acetate had virtually the same boiling point; and that organic compounds in homologous series increased in boiling point by about 19° for each increase in CH_2. These generalization proved to be crucial in helping chemists decide (for example) whether the stable gaseous substance known in the 1850s as "ethyl" had the atomic-weight monomer formula C_2H_5, as Kolbe thought, or the dimer formula C_4H_{10}, as claimed by Gerhardt and Williamson.

The sources of Kopp's interest in these matters are not difficult to trace, nor is the (rarely expressed) *ultimate* goal of his investigations. His fundamental ambition was to shine a bright mental light on the atomic-molecular composition of the chemical compounds he was studying, for his *physical* stoichiometry might well be at least as valuable for the construction of atomic theory as *chemical* stoichiometry had been for Dalton and Berzelius.

It is probable that he had adopted this goal and these tactics from at least two sources. One obvious source was the influence of his professor of physics at Heidelberg, G. W. Muncke. Muncke's *System der atomistischen Physik* (1809) sought to parry the determined attack on atomism in Germany by advocates of a dynamical metaphysics derived from the Kantian tradition. Part of Muncke's investigative strategy was to study, as Kopp was to do, the boiling points and densities of liquids, in order indirectly to enter the hidden world of atoms and molecules.

Parallel to Muncke's work, the law of atomic heats, investigations into the isomorphism of crystals, and early studies of regularities in vapor densities, all of which were published when Kopp was a child, each demonstrated that empirical correlations of physical properties could be crucial for the construction of atomic theories, for all of these correlations were used, in different ways, to assist in the determination of atomic weights and molecular formulae. Such examples from then-recent history surely constituted a second source of Kopp's interest in physical methodology.

In the early nineteenth century the more usual inferential pathway to atomic weights and molecular formulae had derived from pure chemistry. The supreme example of this approach was to be found in the work of Berzelius, culminating in the Swede's final revision of atomic weights and molecular formulae in 1826. However, if one is interested in inference to the microworld, the study of physical properties of substances has one important epistemological advantage over the study of chemical reactions—an advantage that Kopp regarded as decisive. Namely, physical properties reflect a stable substance being observed statically. Chemical properties, by contrast, are determined by chemical reactions, which are dynamic events in which the object of interest is being torn up and reconstituted. How can one draw secure conclusions about what a molecule looks like from evidence derived from its destruction? To be sure, one could never say *how much* of the molecular constitution was altered in the course of a chemical reaction, but before the arrival of valence theory the more natural assumption was for greater rather than lesser scrambling of the atoms as a result of chemical reaction. Consequently, the physical approach to molecular constitutions seemed to be a more certain and reliable route to such knowledge.

Kopp expressed this epistemological admonition throughout his career. As early as 1843, at the end of the first volume of his history of chemistry, Kopp prophesied that pure chemistry would soon be thoroughly united with mathematics and physics, inaugurating a new heroic age of discovery. One could already see this new age beginning to dawn, he thought, and the future looked bright indeed [24, 1:447–55]. Kopp argued along the same lines in the portions of Wöhler's textbook for which he was responsible, and in his treatise on theoretical chemistry: it is by physical and not just chemical means that we can unveil the secrets of the microworld [17, 255–58, 323–44]. And he sang the same song in his lectures.

All this can be very briefly stated: what Kopp was passionately seeking was a way to see, with the mind's eye, the hidden world of atoms and molecules, and to do so with epistemological confidence.

However, there was a contradictory impulse almost hard-wired, as it were, into Kopp's intellectual circuitry. As a student in Heidelberg, Kopp took courses not only with the atomistic physicist Muncke, but also with the university's renowned chemistry professor, Leopold Gmelin. Though it would be an exaggeration to suggest that Gmelin was inimical to theory, he was famously circumspect. It was Gmelin who succeeded in persuading Liebig, Wöhler, Heinrich Rose, and Gustav Magnus that they all should cease mucking about with this or that system of hypothetical atomic weights, and instead immediately return to the system of so-called "equivalents," which Gmelin and some others regarded as the only practical and purely empirical option. This five-person conversation happened during a coach journey in September 1838, a month before Kopp earned his doctoral degree [25, 96–98, 382–86]. This was also an era of growing influence of positivism, a movement being promoted just then by the publication of Auguste Comte's multivolume treatise on "positive philosophy."

At this time, the positivist impulse was often associated with a certain methodology in science, or at least a certain methodological *rhetoric*, namely the inductive approach made famous by the influence of Francis Bacon. We see this influence in a notable polemic in which the young Kopp engaged with H. G. F. Schröder (1810–1885) over the study of molecular volumes. In 1843, Schröder had incautiously implied that Kopp had plagiarized some of his earlier work. Kopp responded with a monographic defense, which included a sharp attack on Schröder's method:

In my work I have always followed the path of specifying: what is the goal of the work in general? How do we proceed? How do we test our results? ... Everything is to be understood by the rational mind, without the assistance of other powers of the spirit. Herr Prof. Schröder disagrees. ... He states what he intends to find right at the start, so that the reader should get into the right frame of mind, "in order to give the imagination something to go on" ... Good God, imagination and a specific weight! In my work I have always followed what has been called the inductive method, which seeks to ground the truths of nature by beginning with the recognition of specific instances, and seeks by continued generalization to rise to general truths. This method is, to be sure, very sober; imagination has nothing to do with it ... [26, 12–13]

Given this inductivist-empiricist mindset, we can now understand even more clearly the basis for Kopp's distrust of chemical evidence for molecular theory. But chemical methods were to prove to be a far more powerful probe into that world than Kopp had foreseen. To put it plainly, Kopp's prediction of a crucial or even exclusive role for physical chemistry proved to be mistaken.

What happened, summarized briefly, is this. Those who would have wanted to see by inference into the interior of chemical molecules were frustrated by the multiplicity of systems of atomic weights and molecular formulae that competed with each other all through the first half of the century. In such a context, the attractions of a more physical approach to the subject were obvious, and we can understand Gmelin's advocacy for the putatively empirical system known as "equivalents," as well as Kopp's preference for physical chemistry. However, beginning in 1850, Alexander Williamson and those who followed his lead forged a new and purely chemical pathway into the intellectual thickets. Williamson and his friends showed that it was indeed possible to provide compelling chemical arguments for one single system of atomic weights, a simple modification of what Berzelius and Gerhardt had each advocated in slightly different forms. This was the heart of the reform movement culminating in the famous 1860 conference in Karlsruhe, whose surprise hero was the Italian Stanislao Cannizzaro. Significantly, not only was this an essentially *chemical* approach; the breakthrough was made possible only by the progress of *organic* chemistry, perhaps a priori the least likely subdiscipline for such agency.

I do not want to oversimplify. Williamson, Kekulé, and Cannizzaro all made important use of physical approaches when it seemed helpful to do so. Indeed, the final step in this process was Cannizzaro's conviction that certain physical methods, such as atomic heats and vapor densities, must sometimes take pride of place over pure chemistry in the final decisions over assignment of atomic weights. But the crucial *new* evidence at the heart of the so-called "quiet revolution" was entirely based on carefully designed chemical reactions, not novel physical–chemical studies.

Kopp was no doubt surprised by this turn of events, but he recognized the significance of what was happening. And in fact Kopp and his physical methods did indeed play a role in the story. The most obvious and striking example of this was Kopp's boiling-point studies of various organic compounds, which provided important confirming evidence, as noted above, in support of the revised molecular formulae that Williamson had defended in 1850, and against those of Liebig, Gmelin, and Kolbe.

In his annual reports on the chemical literature, Kopp reviewed Williamson's first papers favorably but noncommittally [27]. Better evidence of Kopp's personal conversion from Liebig's ideas to those of Williamson comes from publications in 1854, though Kopp expressed himself cautiously, as he always did publicly [28]. In the first edition of Kopp's theoretical chemistry textbook of 1857, he described all the new research of Gerhardt and Williamson, but continued to use the older formulations. He averred there that "recent considerations" (by which Kopp meant the newly revised atomic weights and molecular formulae) had not yet "consolidated," and were not yet so widely adopted that he felt it appropriate to provide a full textbook presentation [15, 676–79, 726–29, 844–60]. But by the time of the second edition, six years later, Kopp had adopted the new formulations, now declaring that they were neither merely conventional, nor arbitrary [16, 352–57]. In his later historical writing, Kopp retrospectively saw Williamson's first ether theory paper of 1850 as a revolutionary turning point in the modern history of the science [29, 736–40, 750–53, 802–8].

Kekulé was powerfully influenced by Williamson's ideas, and he took further important steps beyond what his English mentor had accomplished. Kekulé (and virtually simultaneously, his independent Scottish rival Archibald Couper) showed how carbon atoms could link together to form a "skeleton" or "chain" at the heart of the organic molecule, to which other atoms such as hydrogen, oxygen, and nitrogen could attach themselves according to known valence rules. These papers inaugurated an exciting and fast-moving period in the history of chemistry during which many of the mysteries of molecular structures were gradually resolved. Kopp was not an active participant in this story, but he was a highly interested observer.

Kopp had good reason to exercise caution vis-à-vis these ideas of his brash former student. For one thing, Kekulé's theory was far from perfect, for it was applicable in a fully satisfactory way initially only to a small number of compounds. Furthermore, in the late 1850s and 1860s, structure theory was not the only proposed way to understand the constitutions of organic molecules. An alternative theory had been developed, largely independently, at just the same time, by Kopp's old friend Hermann Kolbe.

Kolbe's theory was based on the radical theory elaborated by Berzelius, Liebig, Bunsen, and others. Kolbe denied that atoms could link up with each other using atom-to-atom bonds. Instead, he posited radicals of various sizes in a strict molecular hierarchy, each radical cohering *as a whole* in the molecule, using isotropic forces (implicitly, at least, these forces were coulombic) [8, 30, 166–69, 174–80]. To add to Kopp's quandary, Kolbe had a fiery, combative personality, whose intensity only increased during the 1860s and 1870s. His public attacks on his scientific enemies, especially on Kekulé, ultimately became so personal, so crude, so violent, and so continuous as to make some wonder whether he was suffering a mental illness. Although Kopp and Kolbe were close friends (Duzfreunde), Kopp could never be confident that Kolbe would not some day turn on him, too, given the right circumstances.

It was no secret, to Kolbe or anyone else, that Kopp was inclined toward Kolbe's scientific enemies, but since Kopp continued to exercise caution in his public positions, the friendship survived. In his discussion of recent history of chemistry at the conclusion of *Entwickelung der Chemie in der neueren Zeit* (1873), Kopp provided a perceptive analysis of the rise of valence and structure theory up to the early 1860s. The chapter is worth reading even today. There had been disagreement on the degree to which such depictions of intramolecular atomic arrangements could be trusted, Kopp noted, "but as the theory was further developed, the confidence of many [chemists] grew, and quickly." He described Kolbe's theory, too, but in the end he demurred further characterizing it, by noting that such matters are of such recent date that they belong not in a history, but in a textbook [29, 824–32].

But Kopp hesitated to use the structural formulae that the structural chemists began using in the 1860s; in fact, virtually no important chemist older than Kopp adopted them. Liebig, Wöhler, and Bunsen were particularly dismayed by the rapid proliferation of structural formulae. These older chemists agreed with Kolbe in viewing them as fantastic products of over-enthusiastic imaginations, with little connection to empirical chemical science. Wöhler reflected this aversion when, after remarking how "cute" it was of Kopp in the *Molecular-Welt* fantasy to endow his personified atoms with little hands, he then added the disdainful remark: "Someday I'm sure we will see the young construction-chemists painting their formulae with little hands on them." He then revealed a certain degree of personal insecurity regarding this subject: "But I can easily believe that I am underestimating the seriousness of the matter and your ingenious ideas and consistent concepts that underlie the whole."[29]

One of Kopp's new literary projects of the 1870s was a thorough analysis of these "ideas and concepts" in chemical theory since about 1850, essentially an updated and expanded version of what he had twice written for successive editions of Vieweg's *Ausführliches Lehrbuch* project, and what he had summarized historically at the end of his *Entwickelung der Chemie in der neueren Zeit*. This new project originated in Kopp's endeavor, starting in the late fall of 1872, to write up in a connected way his oral lectures for his course on theoretical chemistry, which would then constitute a new edition of his book on theoretical chemistry. He worked at this task on and off for four years, often complaining to Wöhler and to Hofmann how difficult the work was.[30]

---

[29] "Wir werden es nur erleben, daß die jungen Constructions-Chemiker ihre Formeln mit Händchen malen. Doch [ich] glaube leicht, daß ich den Ernst der Sache und Deine geistreichen Ideen und consequenten Vorstellungen, die dem Ganzen zu Grunde liegen, verkenne." Wöhler to Kopp, 11 April 1882, MPGA.

[30] Kopp described the origin of the project in a letter to Hofmann, 25 January 1873, BBAW-CB. "Ich arbeite diesen Winter meine Vorlesungen über Theoretische Chemie einmal ganz vollständig aus … [D]as macht mir viele Arbeit … und wie schwer es ist, giebt sich und ergiebt sich wohl beim schriftlichen Formuliren." Progress (or, more commonly, non-progress) reports are often found in Kopp's letters to Wöhler, e.g. 11 March, 7 June, 8 October, 3 November, and 2 December 1873, 30 December 1874, 15 February and 15 October 1876 (BBAW-W); and Kopp to Hofmann, 17 July 1873 (BBAW-CB). These reports cease in the fall of 1876.

There followed a long period of silence about the project, until we encounter a final lament in 1882, when Kopp cited this project as his reason for declining Hofmann's request that he collaborate on a new edition of Hofmann's introduction to modern chemistry.

> I cannot agree to this, because I need to put all my effort into bringing to conclusion what I have begun a long time ago. The more general theories of the chemical composition of bodies have occupied me for more than 7 [actually nearly 10] years. I have repeatedly interrupted the writing, usually because of my health. The result of this has been that what I had earlier written has had to be rewritten, not just once but many times. I see that if I want to bring this book to a conclusion I cannot allow anything else to occupy me, not even those things that I have earlier written and could be finished in a relatively short time. And in any case it is going slowly; I have little time for it.[31]

The reference to "general theories of chemical composition" suggests that the sticking point for Kopp was always the organic-chemical part of theoretical chemistry, namely theories of chemical structure. Despite many years of labour, writing and re-writing the manuscript, he could somehow never bring himself to complete this work. It never saw the light of day.

3 The Making of the Molecular World

What Kopp did succeed in finishing, and in a remarkable burst of speed, was a different sort of book, namely his endearing fantasy *Aus der Molecular-Welt*. This work was a surprise gift in honor of Bunsen's 71st birthday (31 March 1882),[32] which Kopp sent to him in Naples, where Bunsen was then vacationing. The premise of the fantasy is that a party of German friends are touring the Naples "aerarium" (an imaginary airy version of the then-new Acquario di Napoli), stopping by each labeled exhibit in turn—these were large glass balloons in which the atoms and molecules are visible—and discussing the marvelous world they see inside.

According to Kopp's preface, the work originated in what appears to have been his very last experimental project, a study of the nature of hydrated salts in solution, carried out in the fall of 1876. Kopp wrote that he had been interested to

---

[31] "Ich kann darauf nicht eingehen, weil ich Alles daran setzen muß, schon lange Begonnenes zu Ende zu führen. Die allgemeineren Lehren über die chemische Zusammensetzung der Körper beschäftigen mich seit mehr wie 7 Jahren. Wiederholt habe ich die Bearbeitung ausgesetzt, meistens durch meinen Befinden dazu veranlaßt. Das hat zur Folge gehabt, daß das früher Bearbeitete nicht etwa nur einmal sondern mehrmals umgearbeitet werden mußte. Ich sehe ein, daß ich nichts Anderes, nicht einmal Solches, was früher bearbeitet mir vorliegt und in verhältnißmäßig kurzer Zeit zum Abschluß gebracht werden könnte, mich beschäftigen lassen darf, will ich dieses Buch zu Ende zu bringen. Damit geht es ohnehin langsam; ich habe wenig Zeit dafür." Kopp to Hofmann, 13 July 1882, BBAW-CB.

[32] Nearly all authorities cite Bunsen's birthdate as 31 March, and this was Kopp's intended date, but in favour of 30 March see Stock [11, xxix–xxx].

investigate whether molecules of certain salts had the same degree of hydration in the state of aqueous solution as they did when in the solid crystalline state. This project was never brought to a conclusion. However, it did start Kopp wondering what these molecules might look like, if one could actually observe them directly. This play of imagination (Kopp wrote) had led him to compose, apparently initially just for his own amusement, what amounted to a draft of the later *Molecular-Welt*. In the fall of 1876 he had had, it would seem, no thought of actually publishing the fantasy. He had put the manuscript aside, and had forgotten about it.

Five and a half years later, early in 1882, when casting about for something to give Bunsen in honor of his birthday, he uncovered the manuscript. He decided to revise and expand it, and print it up as a surprise. Bunsen and Kopp had often traveled together on holiday; they had been together in Naples at least twice, during the Easter holidays of 1875 and 1880. However, in the spring of 1882 Bunsen and their mutual friend the Heidelberg physicist Georg Quincke had gone to Italy without Kopp, who had begged off because of his heavy workload.[33]

The small (105-page) book initially appeared in this private printing ("als Manuscript gedruckt") of apparently just a handful of copies. As a consequence, surviving copies from this printing are extraordinarily rare.[34] The anonymous author was indicated only at the end of the preface as "H.K.," and there is no dedication. In the preface, introduced with the salutation "Lieber B!", the author explained that the work was a birthday present for "B.," currently on holiday in Naples, and that he had decided to print the thing in hopes that it would be more likely to survive the leaky Italian postal system better than a handwritten manuscript would. The preface was dated "im März 1882," but Kopp was still

---

[33] Regarding these earlier trips see Bunsen to Kolbe, 4 June 1875, ADM 3507, and Bunsen to Roscoe, 7 March 1880, ADM 1000. On Bunsen's and Quincke's 1882 Italian holiday (ca. 14 March to 19 April), see Roscoe [13, 91]; Kopp to Hofmann, 5 March 1882, BBAW-CB; and Bunsen to Roscoe, 22 April 1882, ADM 1008.

[34] The imprint of this private printing is "Druck der C. F. Winter'schen Buchdruckerei zu Darmstadt. 1882." C. Friedrich Winter ran the printing shop in Darmstadt for his brother, Carl Winter of Heidelberg, owner of Carl Winters Universitätsbuchhandlung. All of the firm's books were physically produced in Darmstadt, but editions published for commercial sale showed only Carl Winter's name, and the location of the publisher as Heidelberg. There is no entry for this private printing in the massive *National Union Catalogue Pre-1956 Imprints*, nor is it to be found in any of the online catalogs of the major national libraries, nor in the comprehensive contemporary German books-in-print indexes of W. Heinsius or C. G. Kayser. I have been able positively to locate only five extant copies of this private printing, those listed in the library catalogs of Columbia University, the University of Maastricht, the University of Manchester, the Universitätsbibliothek Frankfurt, and the Bunsen-Bibliothek at the Auer von Welsbach Museum in Althofen, Austria. The latter was Bunsen's personal copy; see the illustration (kindly provided by the Auer von Welsbach Museum), with an early cataloger's speculation in pencil as to the identity of the editor or author of the anonymous publication. I am informed that this copy contains no inscription or other handwritten annotations, though the pages have been cut. One of the other four copies had been cataloged under the presumed author Hermann *Kolbe* until I notified the relevant library staff of the mistake. The copy owned by Columbia University has been digitized for the internet; the full text can be viewed on Google Books.

able to mail a printed copy for receipt in Naples in time for Bunsen's birthday at the end of that month; consequently, the printing must have been done very quickly at the beginning of the month.[35] He simultaneously mailed copies to a few of his closest friends, including at least Wöhler, Roscoe, Kekulé, and Kolbe.[36]

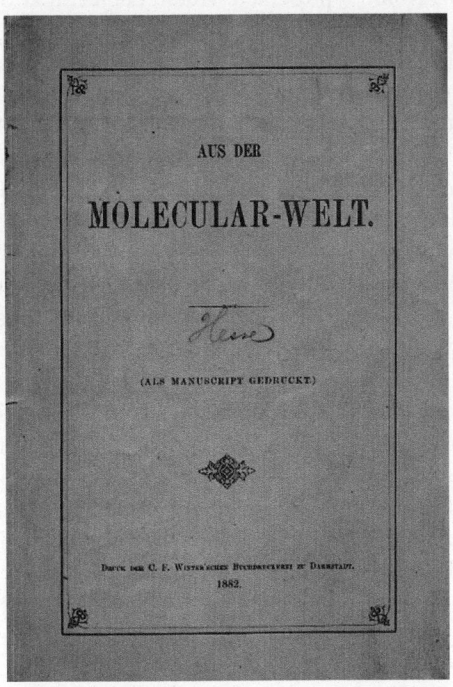

Coincidentally, Kolbe happened to pay an overnight visit to the Kopp home in Heidelberg on 7 March, at the very time that the booklet was being printed; Kolbe was traveling from his home in Leipzig to his customary holiday resort in Gersau, Switzerland. Kopp took this opportunity to tell Kolbe that he was about to publish material that Kolbe should by no means take personally, for it was certainly not meant that way, since he highly valued their friendship. He must have chosen his words carefully, keeping things vague. Five days later, having arrived in Gersau, Kolbe sent Kopp a letter that contained some blistering language regarding Kopp's "mysterious intimations" of 7 March. He had surmised what must have been up.

[35] Kekulé and Wöhler each received copies by the last day of March; see Kekule to Kopp, 31 March 1882, and Wöhler to Kopp, 1 April 1882, MPGA. Remarkably, it was not unusual for both printing and the postal service to be faster in the nineteenth century than is common today.

[36] These four, plus Bunsen, are the only recipients of copies of this private printing for which I have been able to find positive evidence from contemporary documents (Liebig had died nine years earlier). It is likely that Kopp gave Hofmann a copy during the latter's visit to Kopp's home on 16 March 1882 (see below, n. 39). Counting copies for Wöhler, Roscoe, Kekulé, Kolbe, Hofmann, Bunsen, and for Kopp himself, that is a presumed minimum of seven copies printed.

He warned Kopp that Kekulé was an odious thief, and that Hofmann was thoroughly consumed by vanity and by towering ambition. Kopp should beware of associating himself with such scoundrels. Hofmann had surrounded himself in Berlin by sycophants, stooges, and "Judenjungen," Kolbe wrote, and resented the real competition that Kolbe represented. Kolbe conjectured that Hofmann, being too cowardly to act openly, had "commandeered" (gekapert) Kopp, sending him into battle against Kolbe. He concluded with an implicit threat: "I hope I *never* find myself in the position of needing to defend myself against you."[37]

This sort of language was not unusual for Kolbe, and his published tirades were nearly as crude (he had his own journal at his disposal, the *Journal für praktische Chemie*). It is no wonder that Kopp and others stepped carefully around this man. Kopp replied to Kolbe's letter on 16 March. Regarding the publication of which he had spoken and which would appear "very soon," Kopp certainly had not been "commandeered" by anyone; he was far too old to play any such childish games, he wrote. "I have no partisan standpoint whatever, for or against any person, regarding what I say about scientific matters. ... In theoretical chemistry I take care to arrive at a viewpoint, to the best of my judgment, concerning the views that have been brought forward, no matter from whom they derive." He closed by remarking that he had always thought it wise, for the sake of their friendship, not to comment on Kolbe's well-known propensity for personal attacks. But he assured Kolbe that he was dead wrong about Hofmann.[38] Ironically, it seems that Hofmann was visiting Kopp in Heidelberg on the very date of this letter; they may well have conferred on its content.[39]

When Kekulé received Kopp's book in the mail, he sent Kopp a humorously self-mocking thank-you letter. He had been pummeled so often in print by Kolbe that his anxiety was evident.

> I found the booklet this morning in my letter-box, when just starting out, still sober, on my first morning walk. The title appeared suspicious to me, and the signature of the preface heightened my mistrust. H.K. could also signify Heinrich Kolbe, as the Berlin Academy says it. But polemics are always harmful to the nervous State-Hemorrhoidarius, [and] all but poison in the morning, when he is still sober. I nervously shoved the booklet back in the box, to take it on at my leisure, in a stronger mood. Excess of caution! The thing does

---

[37] Kolbe to Kopp, letter draft, 12 March 1882, ADM 3633. I have not seen the letter he actually sent, but one may infer from Kopp's response that its content was either approximately or exactly the same as the draft. For the full text of this letter draft, and a discussion of its context, see Rocke [31].

[38] "Ich habe überhaupt in dem, was ich noch in Beziehung auf Wissenschaftliches äussere, keinen Partei-Standpunkt für oder gegen eine Person. ... Ich gebe mir in der theoretischen Chemie Mühe, eine meiner Beurtheilungskraft entsprechende Ansicht über die Ansichten zu gewinnen, die—sei es von dem oder von Jenem—vorgebracht worden sind." Kopp to Kolbe, 16 March 1882, ADM 3495. Kolbe does not seem to have responded after receiving his copy of the *Molecular-Welt*, and I have found no reference to the book in Kolbe's later correspondence.

[39] Kopp to Hofmann, 13 March 1882, BBAW-CB, looking forward to Hofmann's visit to Kopp's home "this Thursday" (16 March).

good even on a sober stomach, and I have done little else this entire day than treat myself to reading it.[40]

By contrast, upon receipt of the book Wöhler had no thought of Kolbe, no difficulty inferring the correct author, and no doubt concerning to whom "B." referred. He was "surprised, astonished, and admiring" of the "great labour, many-sidedness, and magnificent humor" of the piece. All this without even cutting the pages open, except for the preface![41] Ten days later he wrote Kopp again, saying that he had intended to have read the entire thing by that time, but had gotten through only about a third of it. He had found that the book was by no means light reading, and must actually be studied, for "the rat-king often made the transparency of the sentences unclear at first sight. But forgive this petty remark." The real problem was that the 82-year-old Wöhler was unable to read anything very long before falling asleep.[42] A month later he still had not finished the book, having loaned it to his colleague Wilhelm Weber.[43] Four months after that, he quietly passed away.

4 The Reception

Kopp was probably not candid when he commented in his preface that his reason for having printed the manuscript was simply to guarantee its safe arrival in Italy. He surely must have intended this private printing as a trial balloon preparing the

---

[40] "Ich fand das Werkchen heute früh, als ich, noch nüchtern, meinen ersten Morgenspaziergang antrat im Briefkasten. Der Titel schien mir verdächtig und die Unterschrift der Vorrede erhöhte mein Mißtrauen. H.K. könnte auch Heinrich Kolbe heißen, wie die Berliner Akademie sagt. Polemik aber ist dem nervösen Staats-Hämorrhoidarius stets schädlich, früh morgens, wenn er noch nüchtern ist, gradezu Gift. Aengstlich schob ich das Werkchen wieder in seinen Kasten, um es in kräftigerer Stimmung in Ruhe vorzunehmen. Uebertriebene Vorsicht! Das Ding thut selbst bei nüchternem Magen gut, und ich habe den ganzen Tag kaum etwas anderes getrieben, als mich an der Lektüre zu laben." Kekulé to Kopp, 31 March 1882, MPGA. The reference to sobriety is probably an allusion to Kolbe's frequent public disparagement of "modern" chemists as *not* possessing that necessary quality. The business about *"Heinrich Kolbe"* refers to the circumstance that in 1875 the Prussian Akademie der Wissenschaften misprinted the certificate proclaiming the election of *Hermann* Kolbe to Corresponding Member. Kolbe angrily declined the honor, and returned the document. I cannot offer an explanation of the double meaning of "Staats-Hämorrhoidarius."

[41] "Ich bin überrascht, erstaune und bewundere. Welche Arbeitskraft, welche Vielseitigkeit und welch' prächtiger Humor." Wöhler to Kopp, 1 April 1882, MPGA.

[42] "… ich sehe, daß dieses ergötzliche Product Deines Humors nicht leicht zu lesen ist, sondern daß es studirt werden muß, zumal darin noch manches Mal der Rattenkönig die Durchsichtigkeit der Perioden auf der ersten Blick unklar macht. Doch verzeihe diese kleinliche Bemerkung." Wöhler to Kopp, 11 April 1882, MPGA. For the reference to the "rat-king" see Kopp's preface, below.

[43] Wöhler to Kopp, 8 May 1882, MPGA. This is the last time Wöhler mentioned the book in this correspondence.

way for inevitable commercial publication—as indeed happened. Kopp was under considerable financial pressure at this time, and a private printing of a small book of over a hundred pages was extravagant compared to the alternative of just sending a handwritten manuscript. But it would not have been tactful, considering that the book was framed as a birthday gift to a close friend, to simply rush into commercial publication. He certainly needed the royalty income … but he first allowed others to twist his arm (hence the need for the small number of copies sent to close friends). Kekulé wrote to Kopp (predictably), "But is it really printed as 'manuscript', only for friends? Is it to remain withheld from the larger public? That would be a real shame, and damaging."[44] In a letter to his friend Jacob Volhard shortly thereafter, Kekulé described the privately printed booklet as "wonderful" and "charming," assuming that Kopp had sent a copy also to Volhard.[45] He hadn't.

Kopp made the decision to follow Kekulé's advice (or, more probably, to follow through on his unstated intention all along) and give the work over to Carl Winter's press, the same publisher as for the private printing, for commercial publication. This happened very quickly, for the "second printing" of the book (the first printing for commercial sale) happened within a month or two of the first. Kopp had Winter insert a publisher's note, dated April 1882, that stated that the still-anonymous author "regards it as self-evident that he alone is responsible for what appears and does not appear here."[46] Since the two printings are identical other than the title page and the publisher's note, we know that Winter used the existing plates, carrying out the second printing in probably just a few days. Understandably, Kekulé thought that Kopp's decision to undertake proper publication had probably been prompted by his urging.[47] Four years later a third printing was published, and here for the first time the author's name appears in full

---

[44] "Aber ist es wirklich als 'Manuskript', nur für Freunde gedruckt? Soll es dem größeren Publikum vorenthalten bleiben? Das wäre ernstlich schade, und schädlich." Kekulé to Kopp, 31 March 1882, MPGA.

[45] "Ueber das wundervolle Schriftchen 'Aus der Molecularwelt' hast Du Dich wohl ebenso gefreut wie ich, denn ich nehme an, daß Freund H.K. auch Dir ein Exemplar zugeschickt hat." Kekulé to Volhard, 19 April 1882, AKS.

[46] H[ermann] K[opp], *Aus der Molecular-Welt*, "Zweiter Abdruck," with the imprint "Heidelberg. Carl Winter's Universitätsbuchhandlung. 1882." Like the privately printed edition, surviving copies of this edition are also very scarce (though once again the full text of this printing is readily available on Google Books). The words "Als Manuscript gedruckt" were now omitted from the title page. The publisher's note reads: "Dieses an einen Chemiker von einem Fachgenossen als eine Festgabe gerichteten Schriftchen war zuerst nur als Manuscript gedruckt. Es ist Veranlassung gegeben, dasselbe ungeändert auch in einem größeren Kreise bekannt werden zu lassen. Der Verfasser betrachtet es als selbstverständlich, daß nur er für das darin Stehende und Fehlende verantwortlich ist."

[47] "H.K.'s id est Kopp's reizendes Büchlein 'Aus der Molecularwelt' hast Du inzwischen gelesen. Als ich Dir neulich davon schrieb, wollte ich das sehr durchsichtige incognito nicht enthüllen, obgleich ich an Freund Kopp direkt nach Empfang seiner Sendung einen Dank-Brief geschrieben hatte, der, wie ich glaube, mit dazu beigetragen hat, ihn zu veranlassen, dieses jüngste Kind seiner Laune in den Buchhandel zu bringen." Kekulé to Volhard, 18 June 1882, AKS.

on the title page, along with the full name of the dedicatee.[48] Kolbe had died in the meantime, and there was no longer any reason to be coy.

Kopp's colleagues loved the book. Kekulé wrote Kopp:

> Nothing in ages has given me such pleasure as reading this charming little book. Once again I call it genuine humor, and also genuine science in humorous garb. In our time of constantly recurring swindles, such considerations are decidedly appropriate, and in our time of ultra-personal polemics, the form is actually beneficial.[49]

Wöhler wrote in almost the same terms, characterizing the book as "charming" and "delightful." Roscoe told Kopp, with his characteristic British understatement, that the book "amused me not a little."[50] In his obituary of Kopp published ten years later, Hofmann described the book [10, 519] as "delightful in the highest degree, and perhaps unique of its kind." Similarly, the author of another Kopp obituary [32] characterized the work as an "unforgettable humoresque" and added, "What chemist would not gladly follow him into his 'aerarium' ...?!" An anonymous American reviewer [33] called it "an excellent little work" which "must be read to appreciate its attractive humor and also its fine points of sarcasm." And Thorpe remarked [14, 781], "In the *Molecularwelt*, Kopp's delicate fancy and quaint humour are seen at their best; the book attracted considerable attention even beyond chemical circles ..." Unfortunately, I have found no direct or even indirect information about Bunsen's own opinion of the work.

I have elsewhere argued [30, 279–81] that Kopp's work held considerably more meaning than that of a parody, a mere fantasy, or a simple humoresque. Two contemporaries who shared this opinion, cited above, were Wöhler and Kekulé. Another was Hans Goldschmidt, a prominent student of Bunsen who was twenty-one when the first edition of *Molecular-Welt* appeared, and who thought that one of Kopp's intended didactic targets was Bunsen himself. "This little book," Goldschmidt concluded [34, 2140], "should be highly recommended to every chemist with a sense of humor, showing how humor can be brought even into purely scientific questions." Goldschmidt's, Wöhler's, and Kekulé's opinions are

---

[48] Hermann Kopp, *Aus der Molecular-Welt. Eine Gratulations-Schrift an Robert Bunsen, von Hermann Kopp.* Dritte Ausgabe. Heidelberg. Carl Winter's Universitätsbuchhandlung. 1886." In addition, a dedication page was now added: "Robert Bunsen zu seinem 71. Geburtstag gewidmet," and the publisher's note of the second printing was omitted. "Dritte Ausgabe" is somewhat deceptive in that it implies heavy sale of prior editions, or new revisions; this was on the contrary merely a "Titelausgabe," a new title page simply having been added to unsold copies of the "Zweiter Abdruck." Kopp could not have been pleased with the disappointing total sale of his brainchild. Again, aside from the front matter, the text is exactly the same in all editions, and this edition, too, is on Google Books.

[49] "Seit langer Zeit hat mir Nichts einen solchen Genuß bereitet wie die Lektüre dieses reizenden Schriftchens. Das nenne ich doch wieder einmal ächten Humor und dabei ächte Wissenschaft in humoristischem Gewand. In unserer Periode des nach Willkür wechselnden Schwindels sind solche Betrachtungen entschieden zweckmäßig, und in unserer Zeit der ultra-persönlichen Polemik wirkt solche Form gradezu wohlthuend." Kekulé to Kopp, 31 March 1882, MPGA.

[50] Roscoe to Kopp, 11 April 1882, MPGA.

reinforced by Kopp's prefatory aside, in words addressed to Bunsen {viii}, that "I need to assure you the least of all people, that despite my light-hearted form of expressing myself, I take seriously and highly value" the material treated there. And in another passage later in the book {29}, Kopp wrote: "… I have wanted to discuss all that I have just said in the very serious book with which I have so long been occupied," but which he was now no longer certain would ever get written, because he doubted that he would ever again feel in the mood "to mentally regurgitate the thoughts of the last few minutes."

This "very serious book" undoubtedly refers to his unfinished manuscript on contemporary chemical theories, particularly on the contemporary theories of "constitutions" of molecules, over which he had laboured so hard for so many years, without ever completing the job (see my discussion above, at the end of Sect. 2). I speculate that Kopp may have interwoven certain already-written passages from this manuscript into *Aus der Molecular-Welt*. In fact, it is possible that he had begun to give up hopes of ever completing his history of molecular theory. Incorporating that work into *Molecular-Welt* would accomplish the double task of honoring Bunsen, and simultaneously delivering at least a version of portions of it into the public domain.

This suggestion of a direct relationship between Kopp's unfinished theoretical manuscript and the *Molecular-Welt* is strengthened by two further circumstances. First, in letters to friends Kopp wrote regularly about his work on that manuscript, from its inception in late 1872, until late 1876, the date of the first draft of *Molecular-Welt*, when these references suddenly disappear from his correspondence. This evidence is at least consistent with the suggestion that the book preempted the unfinished manuscript. And second, when Carl Winter sent Kopp 350 marks in honorarium or royalties for the second printing of *Molecular-Welt*, he expressed his hope "that you will now also entrust the scientific MS to the press."[51] I believe that it was this theoretical treatise to which Winter was referring, for there is no other Kopp manuscript, as far as I am aware, that fits the wording of this reference. Winter, at least, appears to have seen a direct relationship between *Molecular-Welt* and Kopp's long-promised but never-delivered theoretical opus. However, Kopp did not accede to Winter's request.

5 Concluding Reflections

Did Kopp's fantasy-world reflect, at least in an indirect fashion, what sometimes went on in his own imagination, when he was operating as a scientist? Kopp suggested exactly this when, in the preface to this work, he traced the origin of the

---

[51] "Anbei empfangen Sie das Honorar für die Molekularwelt mit RM 350. in der Hoffnung, daß Sie nun auch das wissenschftl. Mspt. zur Verlag anvertrauen werden." Winter to Kopp, 17 May 1882, Kopierbuch 1882–85, p. 37, DLA.

piece to a kind of a visual daydream of how the molecules of a salt might look when dissolved in water. Moreover, throughout the *Molecular-Welt* Kopp made repeated appeals to his (and his readers') "mind's eye" (geistiges Auge), the better to see what his personified molecules were doing. He avowed more than once {29, 34} that "such views [Anschauungen] … appeal to me." I suspect that Kopp did indeed want to let his readers gain some insight into his own mental world as a scientist. At a minimum, the work is an appealing glorification of the power of the imagination. It was in this same spirit that Kekulé mentioned Kopp's book in 1890, in the same speech in which he related his two famous "dream" stories about dancing atoms and molecules. Kekulé said [35, 2:942]: "I saw what the venerable [Altmeister] Kopp, my honored teacher and friend, so charmingly depicted for us in his 'Molekularwelt,' but I saw it long before him."

If I am right, then Kopp had come a long way from his avowal as a young man that scientific research must be pursued purely inductively. As he had put it then, "the imagination has nothing to do" with scientific research, which must proceed "soberly," and "without the assistance of other powers of the mind." But if Kopp had followed a tortuous personal path, the science of chemistry had also come a long way in the forty years between these statements. By the latter date, organic chemistry in general and structure theory in specific, as it had been elaborated by Kekulé, Aleksandr Butlerov, Adolf Baeyer, J. H. van't Hoff, Victor Meyer, Emil Fischer, and many others, had celebrated repeated triumphs. This was a branch of the science that had started small in the 1850s, but by 1882 had come to dominate German chemistry to an extraordinary degree, with organic chemistry over-whelming other subdisciplines such as analytical, inorganic, or physical chemistry. And the important point here is that the hallmark of the development of this powerful field had been hypothetico-deductive theory, and not inductive method. In fact, during the middle decades of the nineteenth century the rhetoric and even the reality of scientific method had noticeably shifted from Baconian inductivism toward the method of hypothesis.

To put it simply, by pursuing chemical theory through inductive physical methods from the start of his career, Kopp, like some others, had bet on the wrong horse. Moreover, as a passionate theoretician, a clear-eyed historian of his science, and a thoughtful man, he was wise enough to come to understand this. His inability to bring his late-career theoretical magnum opus to completion, and his trans-formation of portions of that manuscript (as I conjecture here) into a richly imaginative fantasy, is symptomatic of internal psychological conflict—conflict engendered, perhaps, by an intensely theoretical mind having been formed in the same theory-averse intellectual climate that led such of his friends as Bunsen, Heinrich Buff, and Victor Regnault to reject transdictive hypotheses. It is also symptomatic of his failure to break into the very highest scientific rank in his field. He had indeed been caught in the rain of pudding without a spoon.

A final irony of Kopp's career calls for discussion. I have suggested above that physical chemistry remained a marginal discipline throughout most of the nineteenth century, but this statement requires important qualification. Many years ago, J. R. Partington wrote telegraphically but accurately [36, 569], "The name 'physical chemistry' is old, and the subject has always formed an important branch of chemistry." Indeed, investigations that can be brought under this rubric began among the pre-Socratic philosophers, and a continuing thread can be constructed through the centuries. Early nineteenth-century chemists such as Jacob Berzelius, J. L. Gay-Lussac, Michael Faraday, and Thomas Graham made no substantive distinctions between physics and chemistry in their research. And as I have mentioned, physical–chemical research was often important in the construction of chemical theories during the early nineteenth century.

However, during this period such investigations were more often than not viewed as ancillary rather than essential. Mary Jo Nye pointed out some years ago [37, 211] that most German organic chemists of the second half of the nineteenth century had simply left behind the unitary physical–mathematical ideals of their youthful philosophy of science. And as Christoph Meinel observed in his important study of Kopp's colleague and fellow physical chemist Heinruch Buff [38, 259, 263], physical chemistry as usually practiced in the middle decades of the century was more about collecting interesting data than about elaboration of complex theory. Matthias Dörries has explored this subject in connection with the work of Kopp's French contemporary Regnault [39]. In Dörries's view, Regnault's "pitfall of experimental virtuosity" was the "vicious circles" of repetition, pursuit of ever greater exactitude, and expansion of empirical regularities to ever larger numbers of substances, which in some unfortunate cases (such as Regnault's, and arguably Kopp's as well) led to losing sight of the ultimate goal of scientific research.

As I have described it, the "quiet revolution" begun by Williamson and brought to fruition by Kekulé and by Cannizzaro was based essentially on pure chemistry, indeed on organic chemistry, even though physical chemistry played an important supporting role. The heart of this revolution was the devising of techniques by which chemical reactions could provide convincing and epistemically robust inferential access to the microworld, and this access provided the basis for explosive growth in the science after 1850. Such hypothetico-deductive transdiction was largely foreign to physicists in the early nineteenth century. The first continuing tradition of this kind in physics was the kinetic theory of gases, which began to be intensively developed only in the late 1850s.

Physics and physical chemistry did indeed catch up with transdictive organic-chemical methods, but only at the very end of Kopp's career. In fact, Kopp's *Molecular-Welt* appeared just before a crucial turning point in the history of physical chemistry. It is true that Josiah Willard Gibbs had four years earlier completed the publication of his monographic contribution to chemical thermodynamics, but that work was not yet widely known to European scientists. It is also true that by this time kinetic theory was well developed, and its applications to chemical ideas were already part of settled science. But the crucial transition for physical chemistry occurred only after—*just* after—the publication of Kopp's *Molecular-Welt*. Within a few years, physical chemistry had its own revolution, transforming itself, in van't Hoff's colourful language from 1891 [40, 287], from a "colonial possession" to what was now "a great and free country, now being vigorously cultivated."

We need not insist on the magical significance of the year 1887 for the "birth of physical chemistry," as has so often been proclaimed; after all, the first Ordinariat for "physical chemistry" was awarded as early as 1871 to Gustav Wiedemann (to whom the University of Leipzig turned, after Kopp declined this honor) [41]. But it is undoubtedly true that the mid- to late 1880s marked a real transition for this field, setting off a remarkable growth spurt that paralleled that which organic chemistry had experienced a generation earlier. In the last years of the nineteenth and the first years of the twentieth century, Ostwald, Helmholtz, van't Hoff, Arrhenius, Duhem, Planck, Nernst and others fruitfully developed thermodynamics, solution theory, reaction dynamics, electrochemistry, spectroscopy, radiochemistry, and many other areas of modern physical chemistry.

Kopp's discussions of kinetic theory and structural organic chemistry in the *Molecular-Welt* were nothing if not up to date, but his treatments of solution theory and theories of electrolysis, shorn of their imaginative personified character, were little different from what might have been written on these subjects forty years earlier. However, if he had published this same work just ten years later, just before his death, these sections would have appeared embarrassingly old-fashioned. Not only had the factual side of physical chemistry dramatically altered in these ten years, so had the methodology. The new theoretical physical chemistry of the 1890s relied on advanced mathematics well beyond what Kopp commanded, as well as the kind of transdictive and hypothetico-deductive approaches that had long characterized other parts of the chemical enterprise. This is true despite the splash that Ostwald made with his putatively hypothesis-free science of "energetics," which was perhaps less broadly influential than is sometimes supposed [42].

We must also look to broader cultural changes in order to contextualize Kopp's (and his colleagues') intellectual journeys. When Kopp was a young man, the empiricist/inductivist and positivist philosophies were widely praised, at least on the rhetorical level. This may have been part of a negative reaction to the Romantic era, in which certain kinds of metaphysical enthusiasms such as Naturphilosophie were viewed by many leading scholars as having retarded the development of the natural sciences. Public derogations of theory, especially hypothetico-deductive theory, were common. But the cultural climate was different in the last third of the century. Although positivism was still alive and well in certain quarters, celebrations of the scientific imagination were no longer uniformly out of favor.

In his maturity, Kopp was highly self-aware. In the *Molecular-Welt* he added a side comment {**34**}, referring to his lifelong inability to fully commit to views which he found attractive, and which he even believed, but which remained ultimately unproven. "But in the evening of my life, I often find it a bitter thought, that I came to this world with the unfortunate characteristic of constantly seeking my place between two stools." As I have suggested, some part of this ambivalent Hamletian hesitation seems to have derived from his coming of age in a theoretically chaotic period for chemistry, and at a time of positivist and inductivist ideals. Born a generation too early, Kopp had passed through, and cut across, his beloved science. His research had played a principal role in the early formation of the professional discipline we call physical chemistry; but his monumental labours had failed to carry him to his desired destination. It is therefore in many ways most fitting that it is for his superb historical writing, his fine teaching, and his virtuous personal example that Hermann Kopp is best remembered today. And, finally, let us never forget his kind invitation to join him in his fascinating world of molecules.

References to *Introduction*

 1. Buckley A (1878) The Fairy-land of Science. Stanford, London
 2. Meyer LR (1887) Real Fairy Folks, or Fairy-Land of Chemistry. Lothrop, Boston
 3. Gerland O (1893) Geschichte der Familie Kopp. Hessenland 7:171–73, 186–88, 198–99, 244
 4. Siebert K (1919) Hanauer Biographien aus drei Jahrhunderten. Hanauer Geschichtsblätter N.F. 3/4: 108–9, 109–11
 5. Liebig J, Kopp H (eds) (1849) Jahresbericht über die Fortschritte der reinen, pharmaceutischen und technischen Chemie, Physik, Mineralogie und Geologie für 1847 und 1848. Ricker, Giessen
 6. Hofmann AW (ed) (1888) Aus Justus Liebig's und Friedrich Wöhler's Briefwechsel in den Jahren 1829–1873, 3 vols. Vieweg, Braunschweig
 7. Speter M (1938) 'Vater Kopp': Bio-, Biblio- und Psychographisches von und über Hermann Kopp (1817–1892). Osiris 5:392–460
 8. Rocke AJ (1993) The Quiet Revolution: Hermann Kolbe and the Science of Organic Chemistry. University of California Press
 9. Krafft F (1906) Hermann Kopp. In: Weech F, Krieger A. (eds) Badische Biographien 5:406–13. Winter, Heidelberg
10. Hofmann AW (1892) Hermann Kopp. Ber. D. Chem. Ges. 25:505-21
11. Stock C (2007) Robert Wilhelm Bunsens Korrespondenz vor dem Antritt der Heidelberger Professur (1852). Wissenschaftliche Verlagsgesellschaft, Stuttgart
12. Curtius T (1908) Geschichte des chemischen Universitäts-Laboratoriums zu Heidelberg. Rochow, Heidelberg
13. Roscoe HE (1919) Ein Leben der Arbeit. Akademische Verlagsgesellschaft, Leipzig
14. Thorpe TE (1893) The Life-Work of Hermann Kopp. Journal of the Chemical Society 63:775–815
15. Buff H, Zamminer F, Kopp H (1857) Lehrbuch der physikalischen und theoretischen Chemie. Vieweg, Braunschweig. Published also (with second title page) as 3rd ed, vol 1 of Graham-Otto's ausführliches Lehrbuch der Chemie
16. Kopp H (1863) Theoretische Chemie, und Beziehung zwischen chemischen und physikalischen Eigenschaften. Vieweg, Braunschweig. Published also (with second title page) as 2nd ed, pt 2, of Buff, Kopp, and Zamminer, Lehrbuch der physikalischen und theoretischen Chemie, and as 4th ed, vol 1 of Graham-Otto's ausführliches Lehrbuch der Chemie
17. Wöhler F (1868) Grundriss der unorganischen Chemie, 14th ed. Duncker & Humblot, Leipzig
18. Bessmertny B (1932) Hermann Kopp als Chemiker. Archeion 14:62–68
19. Partington, JR (1951) Advanced Treatise on Physical Chemistry, 3 vols. Wiley, New York
20. Kopp H (1839) Ueber die Vorausbestimmung des specifischen Gewichte einiger Klassen chemischer Verbindungen. Ann. Phys. 47:133–52
21. Kopp H (1854) Ueber die specifischen Volume flüssiger Verbindungen. Ann. Chem. 92:1–32

22. Kopp H (1855) Ueber die Abhängigkeit des Siedepunkts und des specifischen Volums flüssiger Verbindungen von der chemischen Zusammensetzung. Ann. Chem. 95:121–26
23. Kopp H (1855–56) Beiträge zur Stöchiometrie der physikalischen Eigenschaften chemischer Verbindungen. Ann. Chem. 96:1–36, 153–85, 303–35, 100:19–38
24. Kopp H (1843–47) Geschichte der Chemie, 4 vols. Vieweg, Braunschweig
25. Rocke AJ (2001) Nationalizing Science: Adolphe Wurtz and the Battle for French Chemistry. MIT Press, Cambridge
26. Kopp H (1844) Bemerkungen zur Volumtheorie, mit specieller Beziehung auf Herrn Prof. Schröder's Schrift. Vieweg, Braunschweig
27. Kopp H (1851–52) Jahresbericht über die Fortschritte der … Chemie 3:459–60, 4: 510–12
28. Kopp H (1854) Ueber die specifischen Volume flüssiger Verbindungen. Ann. Chem. 92:1–32
29. Kopp H (1873) Entwickelung der Chemie in der neueren Zeit. Oldenbourg, Munich
30. Rocke AJ (2010) Image and Reality: Kekulé, Kopp, and the Scientific Imagination. University of Chicago Press
31. Rocke AJ (1990) 'Between Two Stools': Kopp, Kolbe and the History of Chemistry. Bull. Hist. Chem. 9:19-24
32. Anonymous (1892) Chem.-Ztg 16:271
33. Anonymous (1887) Am. J. Phys. 1887:319
34. Goldschmidt H (1911) Erinnerungen an Robert Wilhelm Bunsen. Z. ang. Chem. 24:2137–40
35. Anschütz R (1929) August Kekulé, 2 vols. Verlag Chemie, Berlin
36. Partington JR (1964) A History of Chemistry, vol. 4. London, Macmillan
37. Nye MJ (1992) Physics and Chemistry: Commensurate or Incommensurate Sciences? In: Nye MJ et al. (eds) The Invention of Physical Science. Kluwer, Dordrecht
38. Meinel C (2005) Physikalische Chemie vor 'der' physikalischen Chemie: Heinrich Buff (1805-1878). In: Splinter S et al. (eds) Physica et historia: Festschrift für Andreas Kleinert zum 65. Geburtstag. Deutsche Akademie der Naturforscher Leopoldina, Halle
39. Dörries M (1998) Vicious Circles, or the Pitfalls of Experimental Virtuosity. In: Heidelberger M, Steinle F (eds) Experimental Essays—Versuche zum Experiment. Nomos, Baden–Baden
40. Cohen E (1912) Jacobus Henricus van't Hoff: Sein Leben und Wirken. Akademische Verlagsgesellschaft, Leipzig
41. Barkan D (1992) A Usable Past: Creating Disciplinary Space for Physical Chemistry. In: Nye MJ et al. (eds) The Invention of Physical Science. Kluwer, Dordrecht
42. Hiebert EN (1971) The Energetics Controversy and the New Thermodynamics. In: Roller D (ed) Perspectives in the History of Science and Technology. University of Oklahoma Press, Norman

From the Molecular World

(Printed as Manuscript.)
Printed by C. F[riedrich] Winter's Book-Press
in Darmstadt. 1882[1]

---

[1] This title-page language represents the first (private) printing of this book. See Introduction, Sections 3 and 4, for the printing history of this book. **In Kopp's book, there are no sectionheads (or any other subdivisions); those given in this translation were provided by the present editor, as was everything else that appears here in square brackets. The bolded numbers in curly brackets in this edition indicate the original page numbers of Kopp's book, identical in all three editions.**

A. J. Rocke and H. Kopp, *From the Molecular World*, SpringerBriefs in History of   29
Chemistry, DOI: 10.1007/978-3-642-27416-9_2, © The Author(s) 2012

{v} **[Preface]**

Dear B[unsen]! It sometimes happens that upon the approach of a friend's anniversary celebration one yields to his entreaties, and is induced to promise to do what he can to ensure that this anniversary shall not be celebrated at all, or at least shall be taken as little note of as possible. One then does as little as one can get away with; but the invitations still need to go out, as is necessary for such affairs, and, in the end, one is duly rewarded for his efforts: on the one side, by reproaches for not having sufficiently kept one's promise; and on the other side, by hard feelings that one did not assist with something that others wanted to do, along with the inevitable misinterpretation of motives. This at least has happened to me on one known occasion,[2] and I have taken the lesson from the experience: it is best not to get involved in any way with such an event.

It is a different matter to mark a friend's birthday; one determines this for oneself, or in common with a mutual friend, and others have nothing to say about it. For a number of years past I have {vi} pestered you on that day of the year with the proof that I have remembered it. Several times we have spent this day together in Italy[3]; last year I sent you birthday greetings from that country back to H[eidelberg],[4] and this year I would like to send from the latter city to Naples, where I believe you will be on your birthday, a sign that I have not forgotten the event.

But this time no postcard and also no letter. On this day each year, the memories of our long and close friendship, recalling the melody "Oh Don't You Remember,"[5] resound in me ever more deeply. But at least as far as I am concerned, on this occasion in particular I would like to make the expression of my observance of it a more light-hearted one. But as has happened in many earlier years, so also this year, I have frankly had neither the inclination nor the spare time to dust off and prepare something jocular. You know that lately I have been somewhat more heavily pressed, and have not thought about frivolous amusements.

The thought occurred: perhaps I might have something stashed away that was halfway appropriate? I looked in my "dungeon," namely that compartment of my

---

[2] On 3 November 1881 Bunsen wrote his English friend and former collaborator Henry Roscoe [1, 544]: "I was absent from [Heidelberg] on my anniversary day [the 50th anniversary of the date his doctoral degree was conferred, 17 October 1831], hoping in that way to escape all official notice, but on my return I found so many tokens of kind interest that I scarcely see how it will be possible for me to answer each one separately … and so I am beginning to feel very much exhausted after all I have been through." On 9 January 1882 Kopp wrote Roscoe [2, 91], "We certainly had expected you at Bunsen's celebration. B. had hidden himself away with a few selected friends in Jugenheim on the Bergstrasse [near Darmstadt] … B. bore the unavoidable with dignity and not without pleasure."

[3] Bunsen's surviving correspondence shows that he had made at least two previous trips to Naples on Easter holiday in company with Kopp, in 1875 and 1880; see Bunsen to Kolbe, 4 June 1875, ADM 3507, and Bunsen to Roscoe, 7 March 1880, ADM 1000.

[4] Bunsen mentioned Kopp's Easter 1881 Italian holiday in letters to Roscoe of 5 and 12 March 1881, ADM 1002 and 1003.

[5] "Denkst Du daran, Genosse froher Stunden / Wie wir vereint die Musenstadt begrüsst": a sentimental German student drinking song.

desk into which I throw things that I have started, but whose further elaboration appeared to require more time than I wanted at first to spend on it, or some other obstacle intervened; and also finished articles, if I {vii} was not in the mood to publish them. I do not often call roll for the inhabitants of the dungeon; only in the rather rare cases when it strikes my fancy once more to undertake an investigation begun earlier in the case of one of the inmates—perhaps re-started once or even several times—and maybe this time finally to settle it for good. In the course of such revisions, once in a while something is given over to be printed, many more are neutralized forever, and some are consigned once more to their prison for an indefinite sentence.

Amongst the latter category I found a bundle of papers titled "From the Molecular World." I no longer remembered when it had first been thrown into the jail, which for it really could have become a dungeon; as far as I recall it had never received a hearing in the entire period of its incarceration. Looking through it, it appeared to me that it could possibly be more or less what I was looking for, and then I remembered on what occasion I had written the sketch. When once I occupied myself with the possibility of determining, in a certain fashion for certain salts, whether in aqueous solution they exist as anhydrous salts or as hydrates, and I considered it possible that anhydrous salts unite with water to form hydrates only when they crystallize out of solution, and that in solution the hydrates could be dissociated: {viii} then for the latter case I exercised my imagination regarding what it might look like in a salt solution, how the molecules of the salt and of water would knock about there; from that point the play of my thoughts then turned to gaseous bodies, where we have a stronger foundation for what we know about the molecules of the latter. At that time I did a few experiments for the empirical examination of that possibility, and from my laboratory notebook I could easily fix the time when I pursued this not especially productive subject, since there were very few additional entries, considering the conditions that developed for me in H[eidelberg]; it was in the autumn of the year 1876.[6]

The following pages contain essentially what I wrote at that time. I have gone over it and adapted it to the present purpose, which necessitated some additions. I have also purified it. I have removed some passages which, harmless though they were meant, might have been taken personally. I need to assure you the least of all people, that despite my light-hearted form of expressing myself, I take seriously and highly value many of the advances in our understanding thus described in the following pages. I have untangled the tails of many rat-kings of sentences.[7] At that time even more than earlier I {ix} surrendered to the comfortable temptation of stuffing a great deal in a single sentence, and in this way sentences were formed in

---

[6] The work he speaks of may be the same as that referred to in his letter to Wöhler of 15 October 1876, in which he writes that "das unchemische [?] Arbeiten, dem ich mich mit Liebe wieder einmal hingegeben habe, und das Experimentiren ... ist mir ausgezeichnet bekommen. Es ist zunächst verschiedenes Krystallochemisches, was ich in Untersuchung habe ..." (BBAW-W).

[7] A "Rattenkönig" is the purported accidental knotting of a number of rats together through their filthy tails.

which, as in a set of cardboard boxes, one after another was encapsulated.[8] I have tried to improve at least the grossest examples; Jean Paul says so beautifully: "Do not lose heart when you have erred, and let a more beautiful deed constitute your remorse,"[9] and this admonition should indeed be encouraging, even to those of us who are repeat offenders.[10]

Naturally I later found that the thing cost me somewhat more time, even as a holiday task, than I had at first anticipated, and much more than it was worth. Nonetheless I don't want it to get lost, now that it is on its way to you, and since in Italy letters so often get mislaid, I prefer to send it to you in printed form. Perhaps friend Q[uincke][11] will inquire for general delivery at the Palazzo Gravina[12] at the end of this month, and I will learn after your return that it got to you through him.[13]

With heartfelt congratulations, greetings to Q[uincke], and with the request that you transmit the assurance of my grateful remembrance to Prof. Schr.,[14] Your

H[eidelberg], in March 1882

H[ermann] K[opp]

{1} [Introduction]

In Naples we have often visited the aquarium,[15] and enjoyed there the remarkable animals of the sea-world, along with the fact that there are no monkeys to see there, which are among the most interesting things to see in the aquarium of a Continental metropolis. But why not? for in Munich was there not once a guild ordinance in place that made the commercial sale of suckling pigs the exclusive

---

[8] See the anecdote by Emile Meyerson cited toward the end of Section 1 of the introduction.

[9] "Verzage nicht, wenn Du einmal gefehlt hast, und Deine Reue sei eine schönere That"; Jean Paul Richter (1763–1825) was a German Romantic poet and novelist, known for his wisdom, humor, and aphoristic style.

[10] The reader may agree with the editor (as well as with many of Kopp's contemporaries) that Kopp was not always successful in his simplifying endeavor. See the introduction to this volume for more on Kopp's writing style.

[11] Meant is Bunsen's traveling companion Georg Hermann Quincke (1834–1924), professor of physics at Heidelberg since 1875, and good friend of both Bunsen and Kopp.

[12] A sixteenth-century palatial home which in the nineteenth century was owned by the government, and used to house the Naples postal and telegraph offices.

[13] See Bunsen to Kopp, 22 April 1882, ADM 1008, which describes the trip with Quincke, just concluded. Bunsen did not mention Kopp's present in this letter, nor (as far as I know) in any other surviving document.

[14] "Prof. Schr." is not readily identifiable, but was probably not a travel companion. Wöhler could not guess his identity, either: Wöhler to Kopp, 1 April 1882, MPGA. He may have been one of the "seven permanent naturalists" [3, 86] on the staff of the Naples aquarium (see next footnote).

[15] Completed in 1874 under the direction of German zoologist Anton Dohrn (1840–1909), the Acquario di Napoli is currently the oldest aquarium in Europe. Karl Baedecker, writing in the same year *Molecular-Welt* was published [3, 85], regarded this aquarium, still then under the direction of Dohrn, as "perhaps the most interesting establishment of the kind in the world."

province of poulterers? But today we do not want to visit the aquarium, but an *aerarium* set up by a member of another guild,[16] to see how it goes with the creatures of which airy things actually consist. On our way there we will discuss the question that the name of the place to which we are going can perhaps be judged as incorrectly formed (one of us is actually never sure about this),[17] and that a confusion of the place we seek with one that is devoted to the preservation of more solid matters is near to hand,[18] which confusion however should be charged to those who commit it. But in apothecary-Latin[19] many things happen which linguistically are not entirely in the regulations, and so we will refrain from seeking a better name that could have been applied to the place where we have in fact just arrived.

Using the mind's eye, let us look at some things in the aerarium that especially excite our attention. {2} We direct that eye to any of the variously labeled compartments: how the curious little things are teeming in them.

> Look! Look! By the gallows tree,
> There on the wheel's spindle,
> Half visible in the moonlight,
> Dances an airy mob ...

sang Bürger[20]; but we can notice this sort of mob teeming around not just in one particular place in the air, but everywhere we look. Let us consider what kinds of creatures these are in the light of recent knowledge, and let us express, to those who are not as skilled in the observational art as we, what we here perceive or experience.

[Simple Atoms with Hands]

"Hydrogen": we look inside through the glass that encloses the front of the compartment with this label. What a scramble of incredibly tiny bodies swarming around in pairs. Each of these creatures has but one hand, and this one hand of the first is entwined fast in the one hand of his partner, to form a pair swimming around. This forms as it were a single entity moving around, which one calls a molecule. The creatures that together constitute a hydrogen molecule—such single creatures are called atoms—are, as we said, one-handed (one of our younger colleagues who is not deterred by the coinage of hybrid words would perhaps refer to them as "monomaners"); so each one can seize and hold only one hand of the several held out to him: each has a simple bonding capability. Deprived by nature

---

[16] Presumably a sly reference to academia.

[17] "Aerarium" displays a Latin ending on a Greek root, which seemed incongruous to such a sophisticated classicist such as Kopp. Kopp used the word here with no reference to its meaning in classical Latin as "treasury."

[18] Kopp is presumably referring here to the word "terrarium."

[19] "Apotheker-Latein": i.e., pretentious or incomprehensible formal or technical language.

[20] "Sieh da! sieh da! am Hochgericht / Tanzt um des Rades Spindel / Halb sichtbarlich bei Mondenlicht / Ein luftiges Gesindel." From the ballad "Lenore," by Gottfried Bürger (1748–94).

as these atoms are with regard to *hand*-formation, they are all the more richly endowed with *leg*-power; the molecules they constitute move with fabulous velocity: at $0°$ C. about 1,844 meters in a single second,[21] which comes from the fact that they do not weigh much. These molecules {3} are certainly among the lightest that exist. And even with this lightning-fast motion of the whole molecule, the individual atoms united by this tight handhold feel themselves capable of carrying out a characteristic motion, each for himself; they swing to and fro against one another, rotate around each other, and without thereby causing a diminution of the forward motion of the molecule—as far as we can tell by looking—the one atom slips through under the upraised arms that are bound to the other. It is a refined Ländler[22] that is being danced—respectable, but absent the courtesies [decent, aber rücksichtslos]. Each molecule races in a straight line, until it bumps into another or against the wall or ceiling or floor of the dance-hall, and whenever it bumps, speedily—without even saying "excuse me" after a collision with another, and without doing an about-face—it races just as fast as before, forwards or backwards or diagonally, in another straight line, until it strikes something again and is induced to behave in a similar fashion once more. Within the space placed at its disposal, such a molecule is just like the old everywhere and nowhere; what can resist it, as it restlessly needs to be somewhere else every moment, Leporello's "No rest by day or night,"[23] or *Figaro quà, Figaro là, Figaro su, Figaro giù, pronto prontissimo?*[24]

In another compartment labeled "Chlor" we see generally similarly constituted molecules: pairs of one-handed atoms. The residents of this compartment are actually called "chlorine," but a shortened informal form of the name has come into general use [in the German language]: "chlorine" became "Chlor." These molecules are greenish yellow and have a characteristic smell, but to be sure not like patchouli.[25] They too belong to the Progressive Party,[26] even if they do not progress quite so quickly {4} as the hydrogen molecules do; but to travel a path of about 310 meters every second at $0°$ C. is by no means negligible. But they are also considerably heavier, nearly 36 times as heavy, as the light-footed hydrogen

---

[21] This is the value calculated from theory by Rudolf Clausius [4], in his first paper on the kinetic theory of gases and of heat. This paragraph provides a striking characterization of kinetic theory, as understood in 1882—and as understood today—including vibrational and rotational modes of molecular motion.

[22] An Austrian country dance, characterized by graceful twining of the couple's hands, held securely and continuously.

[23] "Notte e giorno faticar, per chi nulla sa gradir" ("Rest I've none by night or day, scanty fare and doubtful pay"): the beginning of Leporello's opening aria from Mozart's *Don Giovanni.*

[24] "Figaro here, Figaro there, Figaro up, Figaro down, faster and faster": from Rossini's *Il Barbiere di Siviglia.*

[25] An aromatic south Asian plant from which is derived an essential oil much used in perfumery. "Patchouli" comes from a Tamil word meaning "green"—just like "chlorine," from the Greek.

[26] The liberal Fortschrittspartei, arguably the first modern political party in Germany, was formed in Prussia in 1861.

molecules. It is a pleasant arrangement in the molecular world, regarding the requirements for molecular mobility, that a fair consideration is given to their unequal weights, and not all molecules are required to travel at an equal pace; the heavier a molecule is, the (relatively) more comfortably slowly—under similar external circumstances—it is required to run. To be sure, the consideration in this instruction does not go so far that velocities prescribed or permitted to molecules would be nearly inversely proportional to their respective weights, but we also do not go this far among ourselves: in the matter of motion, in the case of an otherwise healthy man who is three times as heavy and who considers himself entitled to move three times as slowly as our nature dictates—that would not suit us. And why should molecules have a more humane arrangement than we people do? So a chlorine molecule, about 36 times as heavy as a hydrogen molecule, may not move 36 times as slowly as the latter, but only about 6 times as slowly. Which in fact it does.[27]

"Oxygen" is written above one of the next compartments of the aerarium. Here we see something different. To form the molecules racing back and forth—they are 16 times as heavy as the hydrogen molecules and move 4 times as slowly—there are once more two oxygen atoms joined together for each, but these are two-handed creatures: equipped with twice as large a gripping or binding power as the hydrogen or chlorine atoms; each of the two oxygen {5} atoms reaches his two hands to the other and forms a double handshake to seal the molecular union. We should immediately ponder the fact that although the molecules of ordinary oxygen gas are constituted in the manner described, the oxygen atoms know how to use their hands in yet another way to form molecules. If electrically excited, three oxygen atoms group themselves to form a single molecule, in which each offers one hand to each of its partners for mutual pressure and mutual bonding; each hand of each of the three atoms is bound in the same way. Oxygen in the latter sort of grouping of its atoms to form molecules is known by the name "ozone"; it behaves with unequal violence compared to ordinary oxygen, and accomplishes effects that the latter is not capable of exerting.[28]

There are other such two-handers like oxygen atoms among the elementary atoms; for instance, sulphur atoms are of this sort. But beyond what nature has considered appropriate to do in the creation of other creatures, she loves diversity regarding the handedness of elementary atoms. For there are also three-handers: in the next compartment, labeled "Nitrogen," the atoms are of this kind, and once again each two nitrogen atoms come together, each offering three hands to the other and taking the three hands of its neighbor, to form one nitrogen molecule,

---

[27] Kopp knew that kinetic energy is proportional to mass times velocity squared, and kinetic-molecular theory as developed since 1857 suggested that average molecular energy was proportional to absolute temperature. So at the same constant temperature, mean (in modern science, root-mean-square) velocities of different molecules must be inversely proportional to the square roots of the respective masses.

[28] That ozone molecules consist of three atoms of oxygen was first proposed in the 1850s; by the 1860s a consensus began to form, and by the 1870s most chemists believed this—e.g., in [5, 234].

and once again such a nitrogen molecule has nothing better to do than to perform the kind of motion that was described for the hydrogen molecule. It cannot appear remarkable that there are also four-handers among the elementary molecules, for corresponding analogies [in the animate world] are found for this; and carbon—which by the way has never successfully been bred in the gaseous state—should be identified as providing an especially {6} important example that the atoms of an element can be four-handed. But five-handedness is somewhat rare, that a single creature (such as an atom of molybdenum or tantalum) has five true gripping hands; for the prehensile tail (*cauda prehensilis*) that some monkeys have in addition to their four hands does not make these animals into five-handers, any more than the possession of the same sort of tail by a quadruped or the possession and dextrous use of the trunk of an elephant makes these animals one-handers.[29] But we must think about the good use of our time, and should not go into this last subject any further, still less should we go further beyond what has been said here.

[The Behavior of Gas Molecules]

We notice that scrupulously tight seals are maintained for all of the compartments of the aerarium, and we discuss how very necessary this measure is, for gas molecules are all too inclined to escape through the narrowest of cracks. One of our party reminds us that at the time when gases were first recognized, it was not understood how to confine them, nor how to condense them to liquids, and it was even viewed as characteristic of gases that both were impossible to do; in the first half of the seventeenth century a certain van Helmont provided the definition of a gas: *Gas est spiritus, qui nec vasis cogi, nec in corpus visibile reduci potest.*[30] Be that as it may, so much is clear, that gas molecules have an irresistible tendency to escape from the space given them; that such a gas molecule, arriving after its zigzag route in its assigned space at the smallest opening in the wall, does not first knock and wait until a voice on the other side cries "come in," but rather straightway rushes in—this is unfortunately {7} all too well known. But the velocity with which it escapes naturally depends on the velocity with which it previously moved through space under the conditions given at the time, and it has already been noted that the velocity of the various molecules vary according to their weights, for heavier molecules move more slowly than lighter ones do. And the variability in the rate at which the heavier and lighter molecules escape through an opening in the wall depends on this factor; for where one can get through, many others can as well.

---

[29] The anonymous authors of the spoof *Berichte der Durstigen Chemischen Gesellschaft*, published in 1886 in Berlin, were surely alluding to this passage when they illustrated the famous benzene molecule formed of six monkeys. The fifth grasping "hand" (prehensile tail) of each carbon / monkey was portrayed as a "residual caudal valence."

[30] "Hunc spiritum, incognitum bactenus, novo nomine Gas voco, qui nec vasis cogi, nec in corpus visibile reduci, nisi extincto prius semine, potest." ("I call this spirit, hitherto unknown, by the new name of 'Gas', which can neither be retained in vessels, nor reduced to a visible form, unless the seed is first extinguished.") Cited and translated in [6, 2:227].

Since in such a case the flight of the various kinds of molecules through the opening does not happen in a manner which with us is considered courteous among adults, first one kind of molecule then perhaps another kind given precedence, but rather takes place in a manner that is difficult to visualize, in order to make it easier for our imagination we have to look not at adults but at children. Picture in your mind an auditorium of a girls' school where a school assembly is taking place on a beautiful day. All the girls are required to attend: from the girls in the earliest classes, with fidgety little bodies that only want to be out in the open and rather running than walking, to the older girls, who, having reached the upper classes, not only take exercises in foreign languages, drawing, music, etc., but also exercises in opening eyes (no longer those that they pursued six years earlier, when they still were knitting their Christmas stockings by following the instruction, "open thirty eyes,"[31] etc.), and who walk slowly and with dignity. There they all sit, paying more or less attention to what is being said. The ceremony drags on. Finally come the words, which for many of our elderly countrymen are the {8} only ones they remember by heart from an important ceremony from their youth: "And I hereby declare this assembly concluded." Everyone stands and moves toward the open doors. Now something happens that is similar to the molecular world. Relative to the numbers of pupils of the first or second kind originally present, in the first few minutes after the doors are opened many more of the small, fast-running kind escape to freedom than of the larger, respectable, more slowly-walking kind. It is exactly the same, when, in the thin wall of a room in which smaller and faster-moving molecules of hydrogen are mixed according to the original proportion with heavier and slower oxygen molecules, a very small door is opened into an empty space; after a short time the door is closed again, and in both rooms a census of those present is undertaken; it is found that in comparison to the heavier molecules, there are a larger number of lighter molecules in the hitherto empty space, and a smaller number of lighter molecules in the first space, than that given by the original ratio.[32]

This sort of thing happens, which has been called (the technical expression must be taken from a foreign language) the effusion of mixed gases. A similar thing, or at least nearly so, also happens when to a mixture of gases not a little door is opened, but rather there is offered as an escape route a wall with many narrow interstices through which nothing but a gas molecule can find its way—a porous membrane. If this membrane is not too very thick, the various different kinds of gas molecules present arrive by diffusion through it into the empty space in almost exactly the same proportion as if there would be in place of the porous membrane a

---

[31] In knitting, the German instruction "schlage dreißig Augen auf" (literally, "open thirty eyes") means "cast on thirty stitches" in English. Kopp is punning between knitting terms used by girls at (say) ten years of age, versus attracting the notice of the opposite sex at sixteen.

[32] The law of "effusion" of mixed gases states that at constant temperature and standard conditions the velocities of flow through a small hole in a thin plate into a vacuum are inversely proportional to the square roots of the densities (hence also of the molecular weights) of the gases. It was first determined, coined, and defined by Thomas Graham [7].

narrow hole in {9} the wall of room that originally held the gases.[33] If the porous membrane is very thick, they get through anyway, but in what proportion?—this turns out to be a distinctly different one than for effusion or diffusion, influenced by something that has been unappetizingly[34] called the transpiration of gases; but to consider this subject would be more complicated than what we would want to dwell on today.[35]

The subject is also not as important as that concerning effusion and diffusion; the latter consideration turns out to be rather simple, as in the case of a space in which molecules of one kind of gas were bouncing around is brought into communication, by means of a narrow opening or by access to a porous membrane, with another space which is populated by molecules of a different kind of gas. There immediately occurs an exchange of courtesies; visits are permitted and received from each space to the other by the different kinds of molecules, naturally with the velocities corresponding to their weights. In this way the population statistics of each of the two spaces constantly change, until they are the same in both; from then on they remain constant, assuming similar external circumstances on all sides, since now in any given time interval just as many molecules of each kind from one space find their way by the aforementioned hole or membrane into the other, as vice versa.

It is of importance for human beings to know what happens in this respect in the airy world of molecules, because something similar happens repeatedly in the human world, even if not in an equally smooth fashion. For instance, the laws of molecular physics just discussed appear to have something to say on the subject of how emigration {10} happens. There are disproportionally many lighter ones, and proportionally fewer ones of greater substance, who do this, and of the few who appear to be of greater substance who go along on the journey, a goodly number of them are now no longer to be numbered among the truly substantial.[36] But the emigration question is a difficult question, because in this subject it is necessary to consider the influence of differing external circumstances, which makes the process less simple, and so we can discuss this serious matter *outside* the aerarium. Now, rather, we prefer to pay attention to what is to be seen *inside* it.

---

[33] This is "diffusion," also defined and characterized mathematically by Grahem [8]. As Kopp correctly states, following Graham, the empirical result is the same as for effusion, though the modern understanding of the mechanism differs in the two cases. The process is described in the next paragraph.

[34] "Transpiration" can be a synonym of "perspiration" both in German and in English.

[35] Once again, Graham studied and named "transpiration" [9, 1:264–70, 900–22], which describes viscous flow of gases through holes which, by contrast to the previous cases, are *not* short—capillary tubes, for example. Mason and Kronstadt point out [10] that there has been much continuing confusion on this subject in the twentieth century; however, there is no confusion in Kopp's treatment.

[36] In this somewhat obscure reference to the demography of German emigration, Kopp used the terms "Leichtere" und "Schwerere," literally, those who are lighter and those who are heavier. He may have been referring to personal wealth.

[Friendship and Affinity]

It is an attractive trait in the character of the elementary atoms that, except under truly unusual circumstances, they do not easily have a hand entirely unoccupied, that is, one that is outside of another that holds it and is held by it; there is something childish in this trait, for similarly a child in its little bed in the evening finds that before feeling ready for sleep he must hold at least one hand in that of his mother's, while sucking the sweet little thumb of the other hand. In order to have no hand unoccupied, so also do the restless atoms, when many of them are simultaneously let free from one another (as often happens), immediately seek a complete exchange of handholds. Satisfaction of such a striving can result in the formation of molecules, when an atom offers its hand or hands to the hand or hands of a similar atom and finds a response. Let us call it "friendship" when two totally similar elementary atoms conclude a molecular union in this way; thus the formation of molecules of hydrogen, chlorine, ordinary oxygen, and nitrogen gas depend in this way, as we have seen, on friendship.

But this trait is not asserted {11} in the same way with all elementary atoms.[37] Over there, whence we feel warmth issuing, is a glass balloon with the label "Mercury Vapor." Uncounted many two-handed mercury atoms are losing their way therein, but in a bad-tempered and rather cumbersome fashion; such a mercury atom is 100 times heavier than a hydrogen molecule, and moves 10 times more slowly. Not one offers a hand to another to follow their zigzag path together as a couple; rather, at each of the perpetual meetings and collisions, each grumbles to the other: "I will not extend my hand to you as a friend, and will not be your brother." So with respect to the satisfaction of the aforementioned trait, what does a mercury atom do, who, as a result of such a morose behavior toward his fellows, has to lead his life "in single blessedness"[38]? Nothing is simpler: he holds one of his own hands with the other.[39]

But by no means do all of the unions that are concluded in the molecular world depend on friendship. By no means are the atoms that bind themselves to create a molecular existence always of the same kind; rather, much more often they are of different kinds. For the latter case what they do in this regard has long been called "affinity." The designation is not appropriate; it derives from a time when people

---

[37] The idea that molecules of gases of different elements consist of *differing* numbers of atoms— oxygen gas molecules being diatomic, for instance, while mercury gas is monatomic—was crucial to the reform of chemical theory associated with Stanislao Cannizzaro at the Karlsruhe Conference in 1860.

[38] Kopp wrote these three words in English; it is a quotation from Shakespeare's *Midsummer Night's Dream*: "But earthlier happy is the rose distill'd, / Than that which withering on the virgin thorn / Grows, lives and dies in single blessedness."

[39] This hypothesis, the internal satisfaction of valence units, was only one of several suggestions invoked at this time to explain the apparent variability of valence—in this case, how normally divalent mercury atoms could behave in the gaseous state as if they had a valence of zero. The same hypothesis was used to account for how carbon atoms, normally tetravalent, could sometimes appear to be divalent, as in carbon monoxide—for which, see Kopp's discussion below, on p. {18}.

were deeply committed to a psychological conception of chemical processes, and the pernicious arrogance of putatively exact knowledge quite nearly resulted in a denial of that sympathy between apparently lifeless creatures which we have already asserted. For so-called chemical affinity is of an entirely different kind from that which can be observed in the human world, different individuals placed close together in amicable relations, or not so amicably standing in each other's way (for instance in matters that have to do with inheritance). It is not the kind of affinity like that of two members of the worthy Swabian folk at the beginning of an acquaintanceship in the good old days, {12} who would seek first and foremost to determine, if otherwise no kinship between them could be determined, whether each could trace himself to a relation, even of the most distant kind, who lived in Kirchheim unter Teck,[40] at which point the matter was entirely clarified, for it was considered common knowledge that all the inhabitants of this town were related to each other, and the basis for a confidential conversation was laid, and sufficient reason established for them mutually to adopt at their next meeting the sort of tone that still occurs, as F. T. Vischer's passages from his work *Lyrische Gänge* testify:

> "My goodness, whom do I see here?" joyfully cries the fellow from Bempflingen;
> "'Tis my cousin, the town treasurer of Bopfingen."
> And the Bempflinger shakes the hand of the worthy Bopfinger
> And the Bopfinger shakes the Bempflinger's hand in return;
> And the Bempflinger then says, "You are surely still the same old merry house?"
> And arm in arm thither the cousins wander.[41]

It is a very different sort of thing when atoms of different kinds clasp hands and wander thither hand in hand, united as a molecule. But as little fitting as the expression affinity is for this, it may well last, for it has become old·habit. Many elementary atoms who behave quite passively with their own kind regarding the attractions of friendship yield to the lure of the so-called affinity and enter into combination to form molecules with different kinds of atoms. Even the mercury atom, which is so cool (I won't use a stronger expression) to other mercury atoms, does not maintain its hard-heartedness. It yields, not perhaps without some inner struggle, for a certain alteration appears to have occurred, signified by a change in colour. Turning pale, with one of its two hands it seizes the one hand of a chlorine atom, with the other the hand of a second chlorine[42]; and when somewhat inflamed [or excited], now turning red [or blushing], it does not disdain grabbing with its

---

[40] A Swabian village, near Stuttgart.

[41] "'Ah, wen seh' ich hier? Herr je! Mein Vetter, der Herr Cameralverwalter aus Bopfingen ist's,' jubelt ein Bempflinger dort, / Und der Bempflinger schüttelt des Bopfingers biedere Rechte / Und der Bopfinger auch schüttelt des Bempflingers Hand. / Und der Bempflinger drauf: 'Du bist doch das alte fidele / Haus noch?' Und Arm in Arm wallen die Vettern dahin." These passages, from Vischer's "An eine Quelle," in *Lyrische Gänge* (1882), affectionately parody Swabian mannerisms; Bempflingen is near Stuttgart, and Bopfingen is closer to Nördlingen. Friedrich Theodor Vischer (1807–1887) was an essayist, philosopher, and poet at the University of Tübingen—and a native Swabian.

[42] Forming mercuric chloride, also called corrosive sublimate, a heavy white powder.

two hands the {13} two hands of an oxygen atom, and so enter into combination with it.[43]

When dissimilar atoms which under reciprocal influence are more readily flammable are brought together, it requires but a slight impulse to dissolve the union created on the basis of friendship and enter into one that rests on so-called affinity. In the dark space over there are brought together equal volumes of chlorine gas and hydrogen gas, hence—as we will soon (p. {69} f.) discuss—likewise equal numbers of chlorine and hydrogen molecules; each of these molecules consists of two identical atoms. As long as everything remains at low temperature in the dark nothing in particular happens; the chlorine molecules move in space just as innocently as if nothing out of the ordinary was in store, and the hydrogen molecules do likewise; each molecule races in a straight line through space with the velocity appropriate to its weight and temperature, until at some point it bumps into something, thereupon rebounding, with the same velocity and at first once more rectilinearly, and there is no noticeable difference whether the collision was with a similar or a dissimilar molecule. But once the space is exposed to bright light, as a result of this external illumination it becomes clear to the atoms in the molecules of the one sort what kind of atoms are together with them in the molecules of the other sort, and now matters turn out differently. Courteously, not even at an accelerated tempo, in a graceful *en avant deux*[44] two hydrogen atoms that are united hand in hand to form a hydrogen molecule approach a pair of chlorine atoms in a chlorine molecule that is similarly floating forward; but when the collision occurs, the result is no longer as happened in the dark, the original couples rebounding away, but rather {14} the two hydrogens unclasp their previous handhold, and the two chlorines do the same, and each hydrogen offers a hand to a chlorine to form a molecular union. Two new molecules are created, on the basis of so-called affinity. With some spectacle, the newly joined couples give their regards to friends and acquaintances and now sign themselves "chlorine hydride" (the chlorine predominates, not just in weight, but also because of its more decisive—between ourselves, we will say more decisively negative—character, while the hydrogen is essentially a more indifferent fellow, a much lighter creature all-around; and, by the way, this occurs in the human world as well, even in cases where the relationship between him and her is very different and occasions no such remark, that the maiden name of the woman is placed before that of the man). Then they float away. They too will give and receive many bumps in the molecular world. After a while we catch sight of such a couple and ask ourselves whether these two have found that enduring inner bliss in their combination which matches the fire with which the union was sealed. Unfortunately we cannot be as convinced of this as we would like to be; the couple appears to be desperately sour, and the

---

[43] "Angefeuert" means either inflamed or excited, "erröthend" means either turning red or blushing—once more punning across the inorganic / human boundary. Mercuric oxide is a heavy bright red powder.

[44] A dance movement in which a couple joined by inside hands steps forward.

cowering hydrogen looks as if he is thinking, "Affinity, oh affinity has brought me to this state."[45]

But we can do nothing to help; so we walk on. A pleasant picture is offered to us there, whence moderate warmth attracts us. We see from the label of the balloon from which the warmth is coming that worthy steam has found a comfortable home here. He has a holiday today; no machine to work, nor is he overheated at the job of transforming tallow to glycerin and stearic acid, or to seize by the arm bodies who with injured wings are attempting with difficulty to fly, {15} accompanying them upwards in the retort, then showing them the way down into the receiver, or to blow things out of vessels that do not want to leave willingly, or to do any of the other myriad tasks he is called upon to perform.[46] Under these circumstances the molecules of water vapor only do what is necessary according to their nature as gas molecules; they move rectilinearly as far as is permitted by the space and by their fellow molecules.

Let us examine one. A two-handed oxygen atom and two one-handed hydrogen atoms are associated together; the oxygen atom holds the hand of a hydrogen atom with each of its two hands. As far as it can, the molecule rushes straight ahead, while each of the three atoms belonging to it performs in addition the motions characteristic to them. The *allemande à trois* is performed in a charmingly beautiful way, nearly as beautifully as it was danced in our youth by the then-famous Alexander Casorti with his two sisters,[47] where also it was particularly the intramolecular movements of the dancers that were so appealing.

When several atoms unite to form a molecule, they usually do this with the expression of energy, and allow the latter to reveal itself as freed. Naturally the thus-formed compound loses just as much energy as becomes free in the act of combination, in comparison to the sum of energies that the atoms individually possessed before they came together. A more particular consideration of this subject would lead us to quite important thoughts, but would require more time than we think suitable at this time and place.[48] So let us resist the temptation.

{16} [Two Conceptions of the Constitutions of Molecules]
The molecules of airy bodies are not always constituted so simply in relation to the number and types of elementary atoms contained within them as has been exhibited in the compartments of the aerarium that we have seen up until now, namely those consisting of at most three atoms and of at most two different kinds

---

[45] The reaction between hydrogen and chlorine only happens in the presence of ultraviolet light. Chlorine is of course (electro−) negative, and hydrogen chloride is acidic, i.e., sour.

[46] That is, tasks elliptically described here: working a steam engine, hydrolyzing fats, carrying out steam distillations, or steam-cleaning vessels of all kinds.

[47] *Allemande à trois, getanzt von Alexander [u. seinen Schwestern] Therese und Victorine Casorti, für Pianoforte zu 2 und 4 Händen* (ca. 1830), in the Staatsbibliothek zu Berlin. I was able to locate no better or fuller reference to Alexander Casorti and his sisters.

[48] Kopp refers here to chemical thermodynamics, in particular free energy changes across chemical reactions, a subject that was only in its earliest development in 1882.

of atoms. A very large number of such molecules consist of three or even more kinds of atoms, or are combinations of more, and often of many more, than three atoms, and consisting of two kinds (occasionally of one) or even of several kinds of atoms. The joining together of the molecule is then no longer as simple and unproblematical as in the previously considered cases. How does this happen?

(We would like to express ourselves on this subject in a very clear way, for among those present with us in the aerarium today are some whom we have repeatedly seen come close by, and have appeared to listen in on our conversation; and to be sure we are somewhat vain. Look: there they are again beside us, among them the friendly married couple and the nice intelligent-looking girl, who appears to be their niece or cousin; the slightest hint of a Thuringian dialect leads us to suspect what part of Germany they are from. The two younger gentlemen who are hanging around them probably are doing that not solely on their account, but they have been looking so often at you, and then whispering to each other, that I do believe that you have been recognized.)

How does this happen ... I mean this joining together of a molecule consisting of a larger number of dissimilar atoms? That is a simple question, but it is difficult to give a satisfying answer, especially if it is also to be generally understandable, and if in addition the answer needs to be a totally reliable one, we would perhaps do better not even to try. But we will do so nonetheless, bearing in mind that *in magnis voluisse* etc.[49]

{17} The public law in force for the airy molecular world[50] is given to us in a very satisfactory way by physics, not without the application of hard-to-use integral signs; regarding this material, one can say in good conscience, provided he has understood the results achieved by deep-thinking scientists: it *will* be thus. Few of these results are necessary for us to pay attention to now, and we will mention later (p. {68} ff.) some matters that even for this light conversation we really should not leave untouched. The private law in force for the same world, information concerning which it is the task of chemistry to provide, is on the other hand known with much less certainty[51]; the details are in many respects still controversial, and at best one can say for any one of the theories that apply here: regarding what actually can be confirmed as happening, it *could* be this way. The understanding of that which concerns especially family law [Familienrecht] and everything connected with it is in a still frustrating state; which institutions are known to be decisive for the relationships of the individual molecules—chemists, revealing their inherent weaknesses and incorrectly applying and confounding the technical expressions borrowed from the classical languages, call the question regarding these institutions the "constitutional" question—concerning these matters we have more conjectures than securely achieved insights, and

---

[49] In magnis voluisse sat est: for great tasks, it suffices to have tried.

[50] Kopp is referring here to the kinetic theory of gases and of heat, as interpreted by the newer methods of statistical mechanics, a field that by 1882 was already well developed.

[51] Here Kopp refers to theories of chemical bonding within molecules.

conceptions regarding the very same subject can be very different, ideas on which contradictory conjectures rest, but to each of which cannot be denied the limited recognition that we stated above.

Concerning the so-called constitutions of molecules of more complex composition there now are principally two conceptions standing in opposition to one another,[52] concerning which it would be good to say at least some few meager words {18} towards communicating an understanding regarding in what ways they differ. That is a tricky matter, both because of the intrinsic difficulties of the task and for other reasons as well,[53] and those who deal with it derive little benefit even in the best case. So I enter into the task on my own authority, proceeding according to the best of my understanding and in consideration of the proper boundaries given above. These conceptions can be elucidated for a molecule containing several carbon atoms. It scarcely needs to be emphasized that one example cannot represent all possible cases; but we also have no intention to treat all possible cases in one example.

As we have already had occasion to remark, every carbon atom is four-handed. It can happen for one carbon atom to have all of its hands engaged if two of them are united with the two hands of an oxygen atom, and the other two are united together (which yields a molecule of carbonic oxide gas); or if two of its hands exchange handholds with the two hands of a first, and the other two hands with the two hands of a second oxygen atom (which results in a molecule of carbonic acid gas); or if the carbon atom places its four hands in the four hands of four one-handed hydrogen atoms (what is thus created used to be called a marsh gas molecule, more recently named methane[54] gas); or in still other ways that are now easy to envision. Let us assume that one hydrogen atom separates out from the last-described molecule A, consisting of 1 carbon atom and 4 hydrogen atoms, in order to enter into a compound elsewhere that it likes better. Now one hand of the carbon atom is free and wants to be accommodated again. That can happen by the offer of the hand of an individual: of a one-handed atom, if such a creature is close by, but {19} also by the offer of a hand on the part of a company business: a group of atoms bound together, which was previously at full strength for pursuing their molecular business, but now however has one hand available as the result of the departure of a one-handed partner in the firm; according to the relevant regulations of the molecular world, the atomic group entering into the molecule A is formally no longer entitled to rights given the hydrogen atom whose place it entered,[55] even if the individual addition of one or another atom belonging to the group can influence how the molecule formerly known as A, as it now has become,

---

[52] Kopp is referring to the opposing theoretical research traditions led respectively by Hermann Kolbe and by August Kekulé.

[53] Namely, Kopp's close friendship with Kolbe, coupled with the latter's well-known pathologically quarrelsome personality.

[54] The names methane, ethane, propane, butane, etc. for paraffinic hydrocarbons were first proposed by A. W. Hofmann in 1866 [11, 57–58n.].

[55] An important qualification in this context. See Kopp's next paragraph for a clearer statement of the same idea.

behaves in certain circumstances. Such an atomic group can also enter into the place of a second hydrogen atom, and this can even go still further. Such an atomic group B can, for its part, also contain a carbon atom. And once more in the atomic group B a one-handed atom previously contained therein can be replaced by a group C that offers instead of its one hand—in place of a woman who properly speaking belonged to group B—there steps, formally with the same authority that the woman had, a group C consisting of one of her relations and the relation's family, a group that once more contains a carbon atom; and so on and so forth.

For my purposes all this has relevance for the elucidation of a conception concerning the so-called constitution of more complexly-composed molecules, to which reference must necessarily be made even in so superficial a sketch as the preceding: it emerges perhaps more sharply in this less appropriate and less fitting simile than in the more natural and less repellant comparison of a molecule with a family; but now it will become understandable without further expenditure of words in what sense an adherent of this conception might say {20} that into a family-molecule A, in which a carbon atom stands as it were as the leader, can enter an atomic group B having within it a similar carbon atom as a member of the newly emergent family-molecule, and once more in this latter atomic group an atomic group C likewise containing such a carbon atom, etc.; that each of these atomic groups, as self-contained entities endowed with special authority, is united with the rest to form the resulting molecule, or is contained in the latter; that in one and the same molecule materially similar atoms—in this case carbon atoms—assume a different significance, in a sense a different dignity, depending on their position and thus their influence on the entire molecule or on a larger or smaller atomic group.[56]

Against this conception is placed another, according to which this inequality of dignity of materially similar atoms within the same molecule is simply not recognized. According to this conception, all materially similar atoms are equal in rank; in their behavior they are unequal only according to what is united most closely with them—i.e., in accordance with the first idea we used, what they have in their hand or hands—and also according to those atoms which, through the mediation of the closest multi-handed atoms, are united with them in their neighborhood and thus bring influence to bear on them.[57]

---

[56] These two paragraphs on pp. {18–20} describe the theory of Kolbe, developed from 1857 until his death in 1884, which asserts that carbon compounds are formed from radicals consisting of carbon atoms that are assigned strictly hierarchical or ranked relationships. For a particularly clear statement of this notion, along with an unfavourable comparison to that of his opponents, see Kolbe [12]; also Rocke [13, 166–69, 174–80]. Here Kopp uses the metaphors of hierarchical organization in a business or a family; Kolbe had used metaphors of hierarchy in an autocratic state or a military unit [12] [14].

[57] This paragraph characterizes the viewpoint of the structural theorists, led by August Kekulé. In this theory carbon atoms form chains consisting of links (carbon atoms) that in principle are equal in chemical importance. This theory was also developed from 1857, but unlike Kolbe's theory it steadily gained adherents, until by 1882 it was clearly the dominant view, especially in Germany and England. The *locus classicus* for this theory is Kekulé's *Lehrbuch der organischen Chemie* (Erlangen: Enke), published in parts from 1859 until 1887, but never fully completed.

Which of these two conceptions should we now make our confession of faith? Which of these two conceptions is really orthodox? Neither of us is in the pleasant position of being able to say, with Bishop Berkeley, that what is orthodox is what I say, and heterodoxy is whatever does not agree with that.[58] We have become very reticent to answer this question; {21} during the many years in which we have followed events in the field of chemistry, we have witnessed the proposal of more putatively permanent views on molecular constitutions than France has had real constitutions in the same period. And all of them, with the exception of only a very few quite recent ones, were like the flowers in spring: they bloom and then die. We have learned something from this. We no longer have true enthusiasm for any theoretical doctrine.

But to the two opposed theories our position is not really cool, much less indifferent. We are convinced that our science absolutely needs such theories. We are glad to greet each new one that greatly improves on what the previous theory was able or would be able to do: that the new theory succeeds, and how it succeeds, in considering from a common viewpoint the steadily increasing facts, even if only to show how they are tied together on the basis of a fiction, to place an *aide-mémoire* in this fiction at the disposition of the memory, which makes it possible to derive, even in an instant, facts that are impossible to remember.[59] We admire that quality even in the most one-sided theories. We hope that if, as is possible, a currently accepted theory is replaced by another, the latter truly accomplishes more than its predecessor, and we wish that the latter might prove itself also in another way, namely that the theory of the future should embrace that for which a theory of the past was better than the theory of the present day, as something that from its viewpoint should harmoniously be considered with the new, and not simply to ignore it. We do not deceive ourselves that when something is ignored, something else will then soon have its turn. Every modern (in its own day) theory in our field is unmerciful toward earlier conceptions that were pre-viously respected, not knowing how to assimilate them and judging them simply as damaging and {22} as something about which younger scientists should best learn nothing at all. And to much that was at one time rightly respected, to him who,

---

[58] I have not been able to trace this saying to Bishop George Berkeley (1685–1753).

[59] On 25 May 1882 the physicist and philosopher Ernst Mach spoke on "Die ökonomische Natur der physikalischen Forschung" at the Kaiserliche Akademie der Wissenschaften in Vienna; it was on this occasion that he coined his soon-famous expression *Denkökonomie* as the principal virtue of successful physical theories. See his [15, 186–213]. It is intriguing to consider whether he may have been stimulated by this similar notion in Kopp's book, whose second printing had just appeared in bookshops. Mach also used another metaphor: theories were like the leaves of a deciduous tree, which allow the tree to flourish and grow for a season, but then wither and fall off. This also is similar to Kopp's metaphor of theories as flowers that bloom and then die.

where it seems appropriate, persists in using the language of a theory that is no longer modern, to all this, evil is attributed.[60] We are no longer surprised at this. The change in viewpoints of the secular can happen in similar ways as changes in matters of faith:

> Where sprouts a new faith,
> Loyalty and love are often
> Like an evil weed torn out.[61]

For the two conceptions which were elucidated above in a most superficial way, the sort of modus vivendi that we as a peaceable people would prefer has not developed; they exist in open warfare. Nonetheless they have much in common, and it can probably be said that much that was done in pursuit of each of the conceptions was of utility for the further development of the other. He who even casually examines how the so-called constitution of certain (especially the simpler) compounds are conceived from the one and from the other side, could even perhaps believe that this battle was over little more than when in that time when there were still old-fashioned night-watchmen, two in the same town responsible for the security of their fellow townsmen were feuding, the one singing *"Protect* the fire and also the light" and the other insisting on the refrain *"Preserve* the fire and also the light,"* in all other respects they being in agreement. Gellert, the poet who tells this lovely story, ended the tale with the words:

> From these so very different ways
> In which each quarrelsomely bound the other in song,
> From this 'protect' versus 'preserve'
> Came ridicule, contempt, hatred and vengeance and rage,

{23} and then adds the further remark,

> These watchmen, I hear many cry,
> Persecuted themselves over such trivialities?
> They must have been great fools.
> Gentlemen, cease your talk,
> You too could be so unfortunate!
> Have you learned nothing from so many great people
> Who in highest dudgeon have estranged one another

---

[60] Especially after 1870, Kolbe repeatedly proclaimed, publicly as well as privately, that the structural theory of Kekulé and associates was not only scientifically worthless but harmful, deeply misleading and even corrupting of younger chemists. For his part, Archibald Scott Couper, independent co-proposer of structure theory, opined in his theoretical paper [16] that the earlier ideas of Charles Gerhardt were "a blunder," "false," "pernicious," "absurd," even "vicious." One of many reasons for Kekulé's greater success was the fact that he was rhetorically kinder to his predecessors.

[61] "Keimt ein Glaube neu, / Wird oft Lieb' und Treu / Wie ein böses Unkraut ausgerauft." From "Die Braut von Korinth," by Goethe (1798).

By their learned quarrels
Over words that signify the same thing?[62]

With this observation (which would have come out quite differently if the poet had had reason to believe that neither of the two watchmen sang as is actually reported) Gellert is mocking that which is clearly seen among humanists, not among scientists; if it had applied to the latter, of course, I should not have mentioned this suggestive and perhaps offensive observation. But the point of view that Gellert thus expresses was quite ill-formed and incorrect. These days, whoever would make this sort of judgment would probably be accused of being frivolous. It almost matters less what is said than how it is said, and less what one thinks than whether one thinks it in the correct form; whenever here or there sectarianism is so raised to the heights that religiosity can benefit little or not at all, it certainly must be for a very good reason. The two conceptions I speak of have moreover distinctly different fundamental ideas; I have earlier attempted to clarify that which fundamentally divides them.[63]

Were it now a question of declaring whether the one or the other of the two conceptions that are now the subject of our conversation corresponds to reality, it would perhaps be wisest to abstain from voting, or, if compelled, to {24} vote for neither one. It would be a different matter if there were included on the day's agenda a discussion of the question: which of the two could command a majority for resolving that one would use that one, until one that is better than either was offered and was proven to be better. Then one would leave himself unprotected if one wanted again to do the wisest thing, namely to say nothing. But if one wants to say something he can do this only for himself; to form a sentence using the word "we" or "one" is not allowable. If I were to be forced to give an official opinion— I am glad that this is not the case—it would be at even greater length than I am known for, and interlarded with "on the one hands" and "on the other hands", and to the signature would be added a distinctly written *salvo meliori*.[64] Among which, in addition to the rather long introduction and much that does not really belong to the discussion and more that is provided for what might be called colouration, the following points, and much else, would surely be included.

---

[62] "Aus dieser so verschiednen Art, / An die sich beid im Singen zänklich banden, / Aus dem *verwahrt* und dem *bewahrt* / War Spott, Verachtung, Hass und Rach und Wuth entstanden. / Die Wächter, hör ich viele schrein, / Verfolgten sich um solche Kleinigkeiten? / Das mussten grosse Narren sein. / Ihr Herren! Stellt die Reden ein, / Ihr könntet sonst unglücklich sein! / Wisst ihr denn nichts von so viel grossen Leuten, / Die in gelehrten Streitigkeiten / Um Silben, die gleich viel bedeuten, / Sich mit der grössten Wuth entzweien?" From "Die beiden Wächter" by Christian Fürchtegott Gellert (1715–1769).

[63] This entire paragraph is of course written in gentle irony. It is likely it was chiefly intended as a tactful piece of advice, or even a subtle rebuke, directed to Kopp's obstreperous friend, Kolbe.

[64] Salvo meliori: with due respect for a better opinion.

In favor of the first and rather more fully described conception is the fact that it claims to be derived in a direct way from a doctrine that once was in high repute,[65] through further development and modification of that doctrine, and as the inheritor of this reputation it feels authorized to assert a certain right of legitimacy; but it has been so modified in the process that it can with perhaps some justification be questioned whether it is not in fact a fundamentally new thing,[66] and the validity of claims of succession in chemistry is questionable, especially where the *antiqua probo*[67] is less easily implemented than it once was. How, according to this conception, the articulation of the molecule of a somewhat more complicated compound should happen; how, in doing this, in connection with one or the other of several materially similar atoms, such as carbon atoms, an unequal rank is acknowledged according to its position in the family-unit of the molecule; all this is best exemplified by an approach toward the realization of the *suum* {25} *cuique*.[68]

For it would not be a derogation of the recognition of what is being accomplished in this direction by the same conception to note that the articulation of a family-molecule is rather complicated, and a clear understanding of this is in no way easy for anyone, not even for those who have had much practice in such matters; everything in nature is of course is arranged in the most appropriate way, but not always in a way that seems simplest to us. It can be truly difficult correctly to perceive the relations between similar atoms in the same molecule, for instance between the various carbon atoms in the family-molecule to which we earlier referred for comparison (p. {20}), or even in a more highly branched example; these atoms stand equally in a parental relationship to one another, which can be unequally more intricate and less transparent than those in the just-cited example. Once again that would not form the basis for an objection, for similar things happen in the human world, and this must be recognized and understood; I remember that a layperson was even able by patience and indefatigable reflection to find his way rather well through the older German genealogies, as my friend W.[69] represented them.

---

[65] Referred to here is the electrochemical-dualist theory of Jacob Berzelius. From these ideas came the concept of radicals within chemical compounds, which were considered to be parts of molecules that function integrally across chemical reactions.

[66] There is considerable justice in this observation. Kolbe gradually eliminated so much of the purely electrochemical portion of the Berzelian theory, and added so many new features in response to valence and type theory, that his model for molecular constitutions really did become a new theoretical entity.

[67] Antiqua probo: approval of the old, i.e., the default assumption that older is ipso facto superior.

[68] Suum cuique: to each his own. In other words, Kopp is suggesting that there are no objective or empirical standards for assigning Kolbe's various ranks to the various carbon atoms of an organic compound—a point made at greater length in the immediately following paragraph.

[69] Not identified.

An objection of a very serious kind might be made, namely that the articulation of a molecule and the rank-relationships of the materially similar atoms contained within it can be imagined in a way that corresponds to this conception, but that it cannot be proven that this situation corresponds to reality. In any case, this conception has in its favor the fact than on its basis extremely important predictions were possible regarding how articulated molecules could be formed from certain numbers of specific elementary atoms, and what chemical behavior could be expected from this or from that predicted articulation of these molecules—some of these predictions {26} have been verified, and some up until now have not—and that by using it in several cases a better insight into the relationships of certain molecules to others has been attained than was previously possible.[70]

The second conception, the principal points of which were explained earlier (p. {20}) in a few words, was also not simply found (so to speak) under a cabbage leaf [hinter der Hecke], but can equally boast of a proud ancestry from theories that were at one time much respected, from which it had developed with the contributions of newly emerging knowledge; among this knowledge was also some that was of capital importance also for the development of the last-discussed conception[71] (we cannot of course enter in more detail into the subject of how or with the participation of this or that person each of these conceptions came to be as they are[72]). As is clear from what was said earlier, the assumptions at the heart of the second conception are of the simplest kind: that the chemical character of a molecule is determined by the material nature and number of the atoms contained in it, and by the sort of grouping of the latter by the offering of hands on the part of two of them each, so that no hands remain free, and that every atom of the same material nature has accordingly a certain number of hands. The simplicity of these assumptions cannot in itself be considered to be something that speaks against the admissibility of the conception based on them.[73] To be sure, for this very reason it has been pointed out that blunderers attempt to make use of this conception and thereby come to worthless and foolish results[74]; but to consider this as proof that the conception is totally useless is just as invalid as to judge the contents {27} of

---

[70] Kolbe's early public triumphs from the years 1859–64 were the confirmed predictions of the structures of malic and tartaric acids, and of the existence of isopropyl and tertiary butyl alcohols. But in the later 1860s and 1870s he made dozens of further predictions on the basis of his theory of hierarchical carbon radicals, asserting the probable existence of innumerable isomers that were never found. It was primarily for this reason that his theory lost ground against that of the structuralists such as Kekulé, which successfully predicted the non-existence of some of these same isomers. For details, see Rocke [17:218–30, 338–39].

[71] Kopp refers here to the theory of types of Dumas, Laurent, and Gerhardt (1838–56), which led to conceptions of atomic valence as developed by such figures as Williamson, Odling, Wurtz, Frankland, and Kekulé (1850–58).

[72] The reader will have noticed how very seldom Kopp included any names at all in his book. This was yet another technique to step softly through the minefields of polarized opinion.

[73] A charge frequently made by Kolbe against structural ideas.

[74] Another claim against Kekulé's ideas, made especially by Kolbe.

the Book of Books by the fact that some—and they have been not few and by no means insignificant people—misuse what they have taken and misinterpreted from its contents to derive nonsensical conclusions.

A much weightier objection to recognizing this conception as fully satisfying is the fact that by consistently applying all of the cited assumptions, not all molecules can be imagined how their atoms are grouped and bonded one to the other (and I am afraid that we will need to come back to this question).[75] But by far the majority of gas molecules can. And the same predictions are possible using this conception as those we cited earlier (p. {25}) as recommending the first conception, and experience has confirmed the predictions offered by the second conception in even greater number than those made by the first, especially the circumstance that the second conception declares, in agreement with experience, many things to be impossible that according to the first appear to be possible and would be expected.[76]

If one should want to convey graphically for visualization how the atoms contained in a molecule are to be imagined as grouped, this cannot happen in an easier or more visualizable fashion than by picturing all the atoms lying in a plane. This cannot correspond to what exists in reality, for the atoms, however they might be arranged in a molecule, must be arrayed in more than two dimensions if their number is large. That is definitely an incompleteness, an incorrectness of the pictures by which one is accustomed to represent the idea of the grouping of the atoms in a molecule according to the second conception, and something of which surely every intelligent person who uses the second conception is aware. Exactly the same thing also applies to the {28} pictures by which the ideas corresponding to the first conception are represented. If only we could come as far with molecules as the botanists have with flowers, the true arrangement of whose parts in space they know; how gladly we would want to satisfy ourselves of many things with pictures that would be comparable in some respects to those of well-pressed flowers. And, by the way, the fully justified attempt to arrive at a conception of how the atoms composing a molecule may be or could be arranged in space has recently been taken up once more, and quite noteworthy results have been achieved.[77] But this would not be the place to go further into this subject.

---

[75] See below, pp. {45–51}.

[76] See above, n. 70. For instance, Kolbe confidently and publicly expected to find a second distinct substance isomeric with benzoic acid, and a second distinct substance isomeric with phenol. He searched assiduously for these compounds for years, but never found them. Kekulé's benzene theory predicted that they are impossible (that is, Kekulé claimed on the basis of his theory that there can be but one benzoic acid, and one phenol). See [17, 292–96, 301–5]. Here Kopp is subtly but clearly indicating his belief that the latter theory is empirically superior.

[77] Kopp is referring here to the early work on stereochemistry of J. H. van't Hoff, and its enthusiastic reception by Johannes Wislicenus [18]. Although van't Hoff's first obscurely published paper on this subject appeared as early as 1874, the hypothesis did not begin to gain extensive notice until 1877. Very little public response was apparent between that time and the publication of *Molecular-Welt*, so this favourable early notice by Kopp is a significant statement for its day.

In sum, even considering all that distinguishes them, the two conceptions have much in common, and especially that both are only fictions and cannot claim the glory of teaching us how it really is in nature. Moreover, that both prove themselves to be useful, as far as explaining the relationships of one body to another: from which bodies the former arises, and to which it can be converted, and more generally what can be called the chemical behavior of a body. Whenever we do not have the [empirical] material whose mastery requires the assistance of one of these conceptions, it is better not to use it at all, first because it is quite unnecessary, and second because it is dangerous. For whenever a young scientist is held in thrall to such a conception, he all too easily views it as if it announces revealed truth to him. How bodies can be thought to be constituted is expressed in an entirely admissible way using the one or the other conception, according to what experience suggests of the chemical behavior of the body; but it is very awkward and difficult to make those to whom the chemical hermetica [Geheimwissen] {29} are to be revealed—at least as long as they remain in the lower degrees[78] and are instructed to believe everything that is told them—believe that the factual behavior of bodies must be as it is for the reason that these bodies are constituted in this or that manner, and for those who know the latter, the former follows as something completely obvious. The result is that young scientists easily become indifferent to acquainting themselves sufficiently with that which is certain, namely the facts, and they become narrower in their understanding and their thought, and more arrogant than they should be.

But how one can go chattering away for hours! For aside from the fact that I have wanted to discuss all that I have just said in the very serious book with which I have so long been occupied, and now I do not know whether I will feel like mentally regurgitating the thoughts of the last few minutes[79]; I say, aside from this, we are spending our time in a way that we did not intend.

[Constitutions of Organic Substances]

For we have come to the aerarium for entertainment rather than for serious instruction and have neglected this fact for all too long. Let's catch up with what we have missed in the last few minutes, regarding what there is to see here. First we must accommodate our mind's eye to the mental picture [Anschauung] that corresponds to the second elucidated conception[80]—for convenience, since it is less demanding for us, and also because the idea appeals somewhat better to a sensibility that is still (or perhaps that is once again) childlike.

---

[78] A reference to mystery cults, Freemasonry, or Rosicrucians.

[79] See Introduction for a discussion of Kopp's "very serious book [manuscript]", which never appeared in print.

[80] I.e., Kekulé's structure theory.

In the next room, which a well-informed attendant calls the *Caldario*,[81] we find it in fact somewhat warm, where not just one but all of the balloons—made of the clearest and thinnest possible glass for easier observation—thank you {30} for warning me just in time, when I was about to walk into one!—are brought to a more-or-less high temperature. We are standing in front of the nearest one, no. 1 of the first series of balloons, whose designation we render in our own language [verdeutschen] as "wood spirit" [Holzgeist][82]; one of us takes off his glasses, brings his face close to the uncomfortably hot glass of the balloon, and cries out, "I see something that you don't see"; the other of us puts on his pince-nez and now sees everything there is to see, much better and more completely. And it is lovely to watch what is revealed to view. Molecules are dancing about, each one uniting together a carbon atom, four hydrogen atoms and an oxygen atom; with three of the four hands of the carbon atom are joined three hands of three hydrogen atoms, the fourth hand of the carbon atom holds one of the hands of the oxygen atom, whose second hand holds the single hand of the fourth hydrogen atom. In this chain, each molecule races along in a straight line until it experiences a collision, after which it again always moves straight; and while it tirelessly hurries on forwards as a whole, the atoms inside carry out their own characteristic motions in the most graceful fashion.

In the next balloon, no. 2, whose label informs us that it contains spirit of wine,[83] we see the atoms united into molecules in still longer chains.[84] Two carbon atoms have each one hand entwined; the other three hands of the first hold the three hands of three hydrogen atoms tightly, while of the three free hands of the second carbon atom two hold the two hands of two hydrogen atoms, the third holds one hand of an oxygen atom, whose second hand is placed in the one hand of another hydrogen atom (the sixth). There are additional {31} containers with other "spirits," but we content ourselves with having viewed wood spirit and spirit of wine, for here too the spirits perform their chain dance in their characteristic way.

Since we cannot see everything, and have to rely on ourselves to seek out what is most remarkable (how painful do we find it that the aerarium is not described in our travel guide ... the most significant sight not given **![85]), we turn our back on this series of balloons and pass on to the next series, which according to the collective label holds *Acids*, and whose arrangement leads us to suspect that the

---

[81] A nonce-word in Italian: a hot-house, spa, or steam room. In ancient Latin, a caldarium was a hot bath- or steam-room in a Roman bath, heated by a hypocaust.

[82] Wood alcohol or methanol, CH_3OH.

[83] An archaic name for (ethyl) alcohol, CH_3CH_2OH.

[84] The term "chain" (German "Kette") was first used to refer to a connected group of carbon atoms by Kekulé, in his first paper on benzene theory in 1865. However, the concept behind the coinage, namely a line of linked carbon atoms, each of which had roughly equal chemical significance, was implicit in his "theory of atomicity of the elements," published in 1858.

[85] Karl Baedeker [3, 85], awarded the Naples aquarium two stars, his highest commendation at that time; of course he did not mention the fictional aerarium.

contents of each of these new balloons correspond in some fashion to the ones we have just partially viewed. It's now quite hot; the attendant, who once again just passed by, explained that if the temperature is too low the inhabitants of each of these balloons cannot be held in the requisite order. We remember in fact once having heard that the molecules of many acids behave in an unruly fashion if they are not kept a seemly distance apart from one another, or if several of them gang up together; so the application of this disciplinary measure appears to be justified.[86]

Let's examine at least one or two of these balloons, to see what's inside ... also molecules consisting of atoms arranged in chains, for the most part. No. 1, "Formic Acid": one carbon atom has joined one hand to that of a hydrogen atom, two to those of an oxygen atom, and its fourth to one hand of a second oxygen atom, which holds with its other hand the one hand of a second hydrogen atom. No. 2: "Acetic Acid": two carbon atoms joined by holding one hand each; the three other hands of one of the atoms are grasped by three hydrogen atoms, and of the {32} three hands of the second atom, two are held by one oxygen atom, while the third rests in one hand of a second oxygen atom, whose other hand is grabbed by a fourth hydrogen atom.[87]

We now feel that we've seen enough in this series. Shall we enter the next room? What meets our gaze is a sign on the wall in large letters which is partially in shadow and hard to see. Is the word perhaps *"Ester"*? There must be a painting on the wall, a picture from biblical history?[88] A close examination of such a painting will offer us a diversion. We walk around the heated balloon-stand in front of which we last lingered, and now find ourselves, not in front of a picture, but in front of another stand on which once again balloons have been placed, and from no. 2 on, several in increasing number one over another. No. 1 has a designation that would be rendered in our language [verdeutschen] approximately as "formic wood spirit ether" [ameisensaurer Holzäther].[89] A curious name! But we know that this corresponds to the nomenclature of a well-respected textbook of 40 years ago,[90] and we can imagine what is intended by it. We are not at all put off by this; about the same time we have often enough consulted a handbook

---

[86] Gas molecules of organic acids show more marked deviation from ideal gas behavior than most other molecules, because of hydrogen bonding and intramolecular polarities. These substances approach more closely to ideal gas behavior as temperature rises.

[87] Formic acid, in both modern and Koppian terms, is HCO_2H; acetic acid is CH_3CO_2H.

[88] This word was first coined by Leopold Gmelin in 1848 [19, 4:161], but it took many years for the term to be routinely used for the chemical combination of an organic acid with an alcohol, in preference to the earlier "ether." Esther is of course the eponymous heroine of the Book of Esther in the Hebrew Bible, a popular subject of old master paintings.

[89] The ester formed from the reaction of wood alcohol with formic acid; in modern terms, methyl formate.

[90] Presumably that of Eilhard Mitscherlich [20, 1:271].

providing information over the entire science of the day,[91] in which inter alia one may find out what is known about "acetic sweet earth" [essigsaure Süsserde].[92]

Let's look a little at the contents of balloon no. 1, at least. What? Shouldn't that be acetic acid again—perhaps it is mislabeled? Once again, as with acetic acid, every molecule consists of 2 carbon, 4 hydrogen, and 2 oxygen atoms. But the balloon we are now looking at feels only lukewarm to our finger, and the molecules inside appear {33} to find themselves in entirely normal circumstances, and previously we were told that for acetic acid molecules this is the case only at considerably higher temperatures. We look carefully into the balloon before us and find: indeed, these molecules are different in the manner in which their atoms are grouped, compared to acetic acid molecules. In the molecules we're currently looking at we no longer see two carbon atoms bound together by joining one hand each; rather, one oxygen atom gives a hand to the one carbon, the other hand to the other carbon atom; the first carbon atom uses his three other hands to hold the three hands of three hydrogen atoms, the other carbon atom uses one of its three other hands to hold the fourth hydrogen atom, and the other two to hold the two hands of an oxygen atom.[93]

Here, too, the arrangement of the atoms in each molecule is essentially in the form of a chain. But this is not always the case. Let's go now into the adjoining room, through the door over which the sign *Compartimento Aromatico*[94] is emblazoned. The smell that meets us here is not particularly aromatic[95]; it smells here like it usually does in the x-alley, where one could imagine oneself transported to Montenegro, even though Turk's heads are not displayed, but instead pale dismembered hands appear to be hung from a line strung across a window; there dwells she who advertises herself so often in the papers as an odorless glove washerwoman.[96] Something must be leaking out here. Very much so—for there in the corner is a broken balloon, the result of a collision with an incautious visitor to the aerarium; but already in its place a new one is being installed in the place marked "Benzol," and we can already look inside. Oh! They're dancing Ring {34} Around the Rosie [Ringel–Ringel-Reihe]! There are six carbon atoms in the molecule, which form a molecular ring with their dark little faces looking out; each carbon gives two of its four hands to one, and one hand to the other of its neighboring atoms, and each tows a hydrogen atom behind it with the fourth hand. How quick and nimble they

---

[91] Presumably that of Leopold Gmelin [19].

[92] In modern terms, beryllium acetate (Süsserde was an old German term for beryllia earth).

[93] Methyl formate, the volatilized substance in balloon no. 1, is HCO_2CH_3; it is an isomer of acetic acid, and this is Kopp's didactic point in this paragraph.

[94] "Aromatic Department," where compounds in the benzene family are exhibited.

[95] In 1855 A. W. Hofmann introduced the term "aromatic" to refer not to the olfactory sense, but rather to denote a chemical family of substances derived from benzene—only some of whose derivatives have notable aromas. By 1866 the term had become well established in the chemical community.

[96] Kopp is referring to dry cleaning (chemische Reinigung in contemporary German), which can be done using benzene. These references are obscure, and perhaps relate to earlier travels of Kopp and Bunsen in common.

are, and at the "all fall down" refrain they crouch; even hydrogens do it! Look how they bow down, straighten up, and twirl rhythmically with unbelievable speed, never letting loose of their handholds right, left, and rear, even with that additional motion which the molecule as a whole executes.[97] And this motion, rectilinear as for molecules of all gaseous bodies, happens at considerable velocity; the molecules of benzene vapor race around in space only slightly more slowly than the molecules of chlorine gas.[98]

[A Personal Comment]

As you know, dear friend, such mental pictures appeal to me, and for that reason I like the [structural] conception used here—not to mention several others—and find it useful, for it lends itself to such mental pictures [Anschauungen]. But in the evening of my life, I often find it a bitter thought, that I came to this world with the unfortunate characteristic of constantly seeking my place between two stools, and this also applies to the present matter. For I like the other [i.e., Kolbe's] conception equally, among others, also because it lends itself no less to mental pictures that appeal to me, and my place is decided on neither one side nor on the other, but for the moment between the two.[99]

[In the Circus]

You know how much I like to visit the circus and how much I regret how infrequently I am able to go when it is in town. Above all I am interested in horses and in performances on them. In this sense I can {35} say, in the words of Uhland: This has to do not just with me, but also with our old mutual friend in G[öttingen], who although so advanced in years is mentally still so fresh.[100] Everyone knows that he, to whom our science owes so much, was an excellent horseman in his youth— namely in the first years of the second decade of our century—and was particularly adept at trick riding; on the equestrian track of a family friend he performed an acrobatic act with his brother, on which occasion his brother broke his arm, at

---

[97] Kopp is describing Kekulé's cyclical formula for benzene, C_6H_6, first proposed in 1865, and by 1882 by far the most widely accepted formula for this important compound. He may also be referring to Kekulé's theory, announced in 1872, that the single and double bonds around the ring may be rapidly alternating.

[98] The molecular weight of benzene is 78, that of chlorine gas 73.

[99] Another intentionally disarming comment, directed no doubt primarily to Kolbe; but Kopp wrote truly when he noted that he had a lifelong aversion to fully committing himself to theoretical ideas. See Rocke [21].

[100] That is, the reason for Kopp's attraction to horses has to do not just with his own intrinsic interest in horses, but also with his association of them with Wöhler. I have not succeeded in tracing this reference to the poet Ludwig Uhland (1787–1862). The phrasing is confusing, and upon first reading it confused Wöhler as well, who was clearly intended here, and who wrote Kopp to say that he could not understand how or why the reference applied to him. "Doch ich nehme an, daß ich zu dumm dazu bin." Wöhler to Kopp, 11 April 1882, MPG Archiv.

which point the friend put an end to this kind of exercise.[101] We still correspond regularly, as we have for so many years,[102] and these days we don't have so much material for our letters; so I am happy that a visit to the circus provides such. But if one goes to see the horses and riding performances, one will also get in the bargain the tight-rope walkers and clowns and everything they do.

One sees lovely things there; what especially excites my attention are those who perform chemical acts. Oddly enough, the friend I just spoke about has never appreciated things that rest on speculation.[103] But nothing can be clearer than e.g. the performance of a molecule of that body which earlier came to our attention as formic wood spirit ether (a different term is now used on the label,[104] but we won't talk about that; we know what is meant). On the carpet covering the sand of the arena lies a carbon atom on its back, holding an oxygen and a hydrogen atom in its strong hands, forming with them a self-contained group, as it were, with which is united another group supported by it and balanced higher up, which consists of an oxygen atom and a self-contained group swaying in the air, {36} formed of a carbon atom and three hydrogen atoms. I saw the acetic acid molecule performed: the same powerful carbon atom holds in the air on one side an oxygen atom and a hydrogen suspended from it, on the other side a group consisting of an oxygen atom and a group supported by it, to which a carbon atom and three hydrogen atoms carried by it are united (this second carbon atom, in this production spatially higher than the first, is of course according to rank and accomplishment subordinate to the first carbon atom that bears the entire ensemble). If, in this topmost group, in the part consisting of a carbon and three hydrogen atoms, in place of one hydrogen atom another even loftier group consisting of a carbon atom and three hydrogen atoms is balanced, this results in the performance of the molecule of another acid, propionic acid; and this can be regularly repeated. The process can go very far indeed.[105]

---

[101] Wöhler would have been in his early teens in the period referred to here, ca. 1813; Wöhler's father was a veterinarian, stablemaster, and estate manager for Georg I, Duke of Saxe-Meiningen. Wöhler was surprised in reading this passage, for he had not remembered ever telling Kopp about the incident. However, the story must have been true, for he praised Kopp's "unermeßliches Gedächtniß" in recalling it; Wöhler to Kopp, 11 April 1882, MPG Archiv.

[102] The Archive of the Berlin-Brandenburgische Akademie der Wissenschaften preserves 576 letters exchanged between Kopp and Wöhler, and there are 67 more letters from Wöhler to Kopp preserved at the MPG Archiv. This is by no means all the correspondence between them that once must have existed.

[103] Wöhler was extremely averse to speculative ideas in general, and especially when they evoked controversy.

[104] Namely, methyl formate; see above, n. 89.

[105] This is Kolbe's view, that organic compounds consist of a hierarchical series of carbon radicals. What Kopp is describing is a metaphorical representation of Kolbe's formulations of methyl formate, acetic acid, and propionic acid. The acrobat or strongman lying on his back in the arena represents Kolbe's "fundamental carbon radical" ("Stammradikal" or "Grundradikal"), the foundation of the respective molecule.

But on this point, when the consequences of one or the other of these two conceptual schemes [Anschauungsweisen] begins to be doubtful or to yield improbable outcomes, the views of different observers can diverge. Whereas someone who considers the viewpoint just illustrated to be more correct can find it hard to believe that a chain formation that, as described earlier, depends only on an exchange of handshakes between adjacent atoms in a correspondingly long line can create a single molecule, so also can someone who prefers the other viewpoint judge in just as unfavorable light the idea of a molecule in just as high a series of atoms placed one above the other. The judgments of the two people can differ according to the viewpoints from which each considers the subject; it is a matter of the effects of perspective, {37} according to which lines that are actually the same length appear unequal to the same viewer if he views them lengthwise or crosswise.

When Corty[106] was here at the beginning of last winter with his company and I saw one evening the performance of the acetic acid molecule as I just described, the next morning about noon I was strolling through the circus grounds, just wasting time, as is my wont. I saw a man leaning against the doorpost at the entrance, and I recognized him as the artist who had performed the part of the fundamental carbon atom; he was taking his ease after a rehearsal. He returned my greeting with unfeigned courtesy; we engaged in conversation; the cigar he was enjoying made him talkative; he seemed to be quite a regular guy. He told me that he had also worked for a while at Renz's.[107] I spoke appreciatively how worthily he had held the acetic acid molecule together the previous evening, which ought to be given even higher billing after the preceding equally demanding performance in a single night in Peking. "Tell me," I then asked him, "how is it that the performance of the acetic acid molecule that I have earlier seen at Renz's show was so differently arranged from what we admired yesterday? At that time an oxygen atom lay on its back on the carpet and balanced on one side a hydrogen atom, on the other a group formed of a second oxygen atom, two carbons, and three hydrogens. Which is actually the correct one?" "Sorry, but you've got me there [ich bin überfragt]," he replied (he was Bavarian, from the Aibling[108] area). "The director is always in charge of that, and if one of us were to ask, 'why do you do it this way, and maybe another arrangement might be possible,' {38} there would be unpleasantness [Unannehmlichkeiten]." "I'm sorry," I said, "but I should have known that."[109]

---

[106] Pierre Corty founded the Circus Corty in 1853. By the time of Kopp's book it was one of the largest circus companies in Europe.

[107] Ernst Renz founded this German circus company in 1842.

[108] Aibling is about fifty kilometers southeast of Munich.

[109] This is yet another indirect reference to Kolbe's extreme unpleasantness regarding alternative theories of the constitutions of molecules.

[Radicals]

Such groups of atoms as we have already considered, which in the place of a hydrogen atom of less complicated molecules form more complicated compound molecules, have the ability to unite with elementary atoms, just as the atoms of elements do. A combination of a group consisting of one carbon atom and three hydrogen atoms with one hydrogen atom is exhibited e.g. in the molecule of so-called marsh gas or methane, or a compound of the same group with one chorine atom in the molecule of a body that is gaseous at room temperature, which one can name monochlorinated methane[110] due to its relationship to methane. Groups of atoms that have this ability can be still simpler or still more complicated in their composition (having a greater number of elementary atoms) than those examples we have considered, and instead of uniting with elementary atoms, they can unite also with groups of atoms.

Also endowed with this ability are atomic groupings of a molecule which we earlier considered as formed from an exchange of handshakes between the atoms composing it, by consequence of which results that wherever the handshake is released the hitherto connected hands are loosed; in this way hands become free with which the resulting atomic groups can seize and hold free hands of elementary atoms or of other atomic groups. For a group of atoms that results in this way an expression of a French chemist has often been applied for more than forty years, namely a "residue."[111] The name is not good, in so far as it can lead to {39} a less benevolent interpretation, as if the object referred to is left over in a somewhat contemptible sense. This would be quite inappropriate. Let us therefore speak more respectfully, if we want to refer briefly to such atomic groups, as parts or pieces of the molecules of compounds.

There exist parts or pieces of compounds which, separated from that with which they were earlier united, and united with different things to form compounds, communicate a common behavior to all compounds in which they enter, especially the ability to bring forth one and the same characteristic substance. Such parts or pieces of compounds are considered as radicals.[112]

Among these radicals, all of which according to what we have said have free hands at their disposal to seize and hold other hands, are many that are eager at any cost to make use of this ability, not managing their affairs as adherents of liberal politics and in no way taking as their guideline the spoken word of an Austrian statesman (as the outcome demonstrates, erroneously), "We can wait."[113] These

---

[110] I.e., methyl chloride.

[111] Charles Gerhardt is meant, who introduced the concept in 1839 (the French word is "résidu," the German word "Rest"). See [6,4:417]. In common parlance, both "résidu" and "Rest" can mean remnants, or even leftover discards from a meal (as alluded to in the next sentence).

[112] See above, notes 56, 70, and 105, with associated material.

[113] Dismissive rejoinder of the Austrian minister of state Anton von Schmerling (1805–1893) after the promulgation of the liberal 1862 constitution which he had championed, when Hungarian nobles, offended by Schmerling's obstinacy as well as by his liberalism, refused to send delegates to the new parliament.

are especially the ones that have an uneven number of hands. We consider here only those with one hand, which are among the worst of their kind. These pieces or parts of compounds seek with great energy to be considered and chosen as good parts of the atoms or atomic groups that appeal to them best under the current conditions, and to unite with them. They do this without considering whether the objects of their desire have already entered into other combinations and thus are no longer free; in the most aggressive fashion they assert their {40} right to a hand with respect to all atoms or atomic groups to which they believe they might expect an accommodation, they thereby often create radical disturbances and alterations in the preexistent social relationships of the circle of the molecular world in which they were released and in which they exert their seductive arts. Only in extreme cases, and when they find nothing at all with which to unite in the existing circumstances, do they remember to "join brotherly hands together,"[114] or they make a virtue of extremity, as if truly like gladly associates with like; then do two similar radicals shake hands to form a more or less stable union.

There are e.g. atomic groups of which each consists of a carbon and a nitrogen atom. Of the four hands of the former, three are entwined with the three hands of the latter; the fourth hand that is still free, although strictly speaking belonging to the carbon atom alone, is often designated nearly as the free hand of the latter, because it acts fast in the name and on orders of the atomic group. In chemical processes, whenever such an atomic group previously found in a molecule united with others receives the *consilium abeundi*[115] from the compound, and can seduce a hydrogen atom that hitherto was likewise residing in a compound, it does so, and enters into a combination that has poisonous character; a molecule of prussic acid [Blausäure][116] is formed. The two-handed atom of mercury can in an indirect way at low temperature be made to entwine its two hands with the two hands of two such atomic groups, but at high temperature they separate; then these two atomic groups place their two otherwise free hands in one another and form together a single molecule.[117]

Whatever such atomic groups one deals with, it is all too clear that they are very sly and know how to skillfully behave as if they were {41] quite simple creatures. The atomic groups discussed above that consist of one carbon and one nitrogen atom imitate e.g. chlorine atoms, they unite inter alia with hydrogen or mercury, etc., according to the same numerical ratios as does chlorine. They have picked up the latter's behavior so exactly and know so well how to reproduce it that they even sometimes form compounds whose external appearance is scarcely

---

[114] "Brüder reicht die Hand zum Bunde." Originally a Latin poem used in Masonic ritual, Germanized by Johann Gottfried Hientzsch (1787–1856) and put to a Masonic melody putatively composed by Mozart, the same tune is now used with different lyrics in the Austrian national anthem.

[115] Expulsion order.

[116] I.e., hydrogen cyanide, which is highly poisonous.

[117] Cyanogen, $(CN)_2$.

distinguishable from the corresponding chlorine compounds.[118] To increase the deception, such atomic groups have adopted names that do not betray their compound character, and could probably lead an innocent person to assume that he is dealing with simple atoms. For example, the last-named atomic group repeatedly used as an example of how one-handed radicals behave is called cyanide. Others behave similarly.

The atomic groups that are entitled to have an even number of hands are not quite so dangerous, and are encountered in the molecular world without quite so clearly—or we should say, so disturbingly—revealing the tendency to place their available hands in those of atoms or atomic groups. Their means permit them, to be sure, to satisfy this urge after accommodating each free hand in another in a special way. Let us clarify this with a very simple example. A molecule of ethylene gas consists of two carbon and four hydrogen atoms. When these atoms come together to form a molecule of ethylene, what first appears is one hand of the first and one hand of the second carbon atom; each of these atoms further seizes with two hands the two hands of two hydrogen atoms, and keeps one hand free. One carbon atom says to the other, reaching his free hand toward the free hand of the other, "Take {42} another hand!"; the other shakes on it, and with that all hands are accounted for. But if there is something to seize for which the ethylene molecule has a desire, then one of the two handholds which the two carbon atoms have entered into is released (one bond remains and is sufficient to keep the two carbon atoms together), and with two newly freed hands it grabs for what it desires.[119]

[In the Dance Club]

It is sad but true that the more dangerous acquaintances are in general the more interesting. If making the acquaintance of other one-handed radicals appeals to us, then let's go to that cabinet over there, where in a comfortably heated room a kind of club has been established, in which are admitted as regular members only those molecules whose structures contain creatures which are of the sort of which we speak, and are of a very definite chemical character. This club also has a special designation for the meetings that take place there, which would read in German something like *réunions des esprits*.[120] The purpose of the society: dancing. We are not allowed in; they are very exclusive here. But we can look in through the glass.

---

[118] Cyanogen is still sometimes referred to as a pseudo-halogen.

[119] Kopp is describing olefin addition reactions. The earliest to be discovered was the combination of chlorine gas with ethylene gas, which produces the oily liquid ethylene chloride. This reaction was discovered in 1794 by four Dutch chemists (J. R. Deiman, A. P. van Troostwijk, A. Lauwerenburgh, and N. Bondt). These men with the difficult names soon became known for short as "the Dutch chemists," ethylene chloride as "the Dutch oil," ethylene as "olefiant (oil-forming) gas," and organic substances akin to ethylene as "olefins" (oil-forming substances).

[120] "... welche im Deutschen etwa *Réunions des esprits* lauten würde." It is another inside joke.

They are dancing there, the molecules! Sometimes faster, sometimes less fast, according as they are lighter or heavier, they pass before our curious eye;

> They dance away, sometimes as if on wings of the north wind
> Blown downward with the current,
> Sometimes as if waving in the soft west wind,

hence as if they were Wittekind's bards practicing Tialf's art.[121] Every molecule persists in a rectilinear path until it bumps somewhere, when it rebounds or darts sideways like a rabbit, then once more {43} moving ahead on its straight-line path; each can say of itself: I do not diverge a single finger's breadth from a straight path. They are dancing there, the molecules, each as a whole, but within every molecule, each atomic group (if one looks closer, even within each individual atom, but we won't speak of that yet[122]) is likewise dancing merrily.

Do you recognize these little ladies who so gracefully dance with others to form a molecule? There goes the light Methylie past us, who even in terms of radicals is not yet even of tailor-weight[123]; on her father's side she has good ancestry—to be sure, not porphyrogenic, but at least pyrogenic—but her mother is a wooden creature.[124] There further back we see little Ethylie, and there the heavy Amylie is brushing past; five Methylies taken together weigh scarcely more than one Amylie.[125] We know her, and also a number of her girlfriends of similar character. In our good years they were known only by their first names, and by the fact that they were of a certain sort of radicals; more recently—I think that they were induced to this step in Würzburg—they have taken the family-name *Alkyl*.[126]

---

[121] "Sie tanzen fort, bald wie auf Flügeln des Nords / Den Strom hinunter gestürmt, / Bald wie gewehet von dem sanften Weste." From the 1767 poem "Die Kunst Tialfs, von Wittekinds Barden," an ode to Nordic mythology by Friedrich Gottlob Klopstock (1724–1803).

[122] An apparent allusion to the hypothesis of subatomic particles, for which at this time there was almost no direct evidence. But in letters to Liebig and to Wöhler in 1863, Kopp had engaged in such speculations (Kopp to Liebig, 23 May 1863, BSB; Kopp to Wöhler, 24 May 1863, BBAW-W).

[123] "[Sie] hat noch nicht das Schneider-Gewicht"; old German colloquialism for light in body weight (literally, 100 Pfund).

[124] All of these are word plays referring to the methyl radical, the lightest in the paraffin (alkyl) family. "Porphyrogenic" = born to the purple = of royal blood. "Pyrogenic" and "wooden" both refer to the production of methyl ("wood") alcohol by the destructive distillation of wood. This was first obtained, and named from Greek roots, by Dumas and Peligot in 1834: methyl = methy + hyle ("spirit" + "wood").

[125] The methyl radical has one carbon atom. Ethyl radicals (named by Liebig in 1833 from the Greek and German word "aether" plus Greek "hyle") have two. Amyl radicals ("amilène" was coined by Auguste Cahours in 1840, to designate a substance from potato starch after fermentation and distillation) have five.

[126] "Alkyl" was coined without fanfare by Johannes Wislicenus, professor at Würzburg; an early use (perhaps not the first) is in his 1882 article [22, 244]. The word was derived from the first three letters of "Alkoholradicale" combined with the suffix -yl; it was (and is) a generic term for any of those radicals who bear the "first names" methyl, ethyl, propyl, butyl, amyl, etc.

But what kinds of partners do they have for these dances? Those atomic groups who have obviously taken care to appear proud and unsatisfied? These partners look so much alike that they can be confused for one another; all have the same features, the same nose; they are obviously all of a type.[127] When talking to these conceited fellows one would hear from each, "My name is *Hydroxyl*." But we have a good memory for chemical physiognomies. Are these not the same fellows who under the name *eurhyzene* were so ridiculed publicly nearly forty years ago by a long dead (and properly honored only posthumously) French chemist, to the point that it was said that no dog {44} wanted to take a piece of bread from them?[128] They are. And now under this new name they stand in high reputation in the molecular world.[129] These guys have kept themselves well preserved.

With such partners these ladies live in the sort of relationship where the latter are seen by far the most frequently with the former. There is always plenty of scandal in the molecular world, and the said ladies are subject to it, as are their relationships. Thus it is said that Methylie along with her customary procurer lets herself be used by the authorities to denature the relationship that Ethylie is in.[130] Amylie, who anyway is actually rather disgusting (most of her compounds are in bad odor) and unpopular (even her girlfriends, some of whom are no better than she, call her, but not to her face, "Fusel-Ma'm'selle")—Amylie and her consort, this noble couple, are said to find it so extremely unpleasant that they must hide themselves when they are together with the better Ethylie and her boyfriend, in order that the group as a whole should appear to be palatable [geniessbar]; and if someone spends an evening in their company in order to enjoy Ethylie's witty conversation, Amylie out of spite will so work on him that the next morning he will recall the pleasures that he has enjoyed with ill-defined woe; and since he is ashamed to say that Amylie had been present, he usually conceals his condition as well as he is able.[131]

I do not think it appropriate to reveal what else I have been told about these ladies under a seal of silence by a no-longer-young basic salt molecule with a

---

[127] A double entendre on the chemical word "type," introduced by Dumas in 1839.

[128] This nonce word was Auguste Laurent's coinage [23, 354–55] (apparently from the Greek for "good radical"). He meant it derisorily, to designate the sorts of imaginary pieces of molecules being proposed by his opponents, the advocates of electrochemical-dualist radical theory. Laurent's ideas were much ridiculed in the 1840 s and early 1850 s, but his true significance was ultimately recognized.

[129] The term "hydroxide" emerged first in the 1860s. See Roscoe to Kopp, 29 January 1867, MPGA.

[130] The poisonous methyl alcohol was (and still is) used by governments to denature ethyl alcohol, in order to allow manufacture and sale of the latter for chemical purposes without beverage taxes.

[131] This sentence contains several puns or double entendres. "Ethylie" stands in for ethyl (i.e., grain) alcohol. "Bad odor" = bad reputation, in both languages; several compounds of amyl do have disgusting aromas. Amyl compounds are also constituents of fusel oil, a mixture of undesirable but unavoidable naturally occurring contaminants in grain alcohol, which many believed (and believe) cause hangovers.

fashionable formula-cap, who still maintains only a loose relationship with the blasé water of hydration in which the affinities have also already been fully balanced.[132] {45} Regarding what is told inside and outside of the molecular world, one must be discrete.

[Atoms of Variable Handedness]

We have considered the elementary atoms as entities that can form molecules with similar or dissimilar types by mutual grasping and holding of hands, and we have considered, regarding the handedness or so-called bonding power of atoms (how many hands they have for gripping or bonding of this or that other atom) what the usual behavior of the atoms can be assumed as true or correct. It may appear obvious that it is not required of each atom, when it has already gripped as many hands of other atoms as it possesses (as we just explained), that it should do anything further in the giving or receiving of hands. An atom which is so unreasonably asked to do such a thing could, one would think, simply refuse, and justify its refusal by citing, in the general sense, the *ultra posse nemo obligatur*,[133] or *impossibilium nulla obligatio*,[134] and, in the particular sense, asserting (that which is substantially and essentially identical to the *exceptio caesarea*) the *exceptio cappadociana: qui nil hat* (in the concrete case, no free hands), *nil dat*.[135] An atom cannot be seen as obligated to meet a demand that appears not only so exorbitant, but one of which it is incapable, even if it were to have the will. For example, we have described the chlorine or hydrogen atom as one-handed, and accordingly we naturally imagined that such an atom who has offered his hand to an atom or has had a hand offered to it can seize and hold no other hand, without letting go of the hand it has already taken. We have described the oxygen atom as two-handed, and also thought of the sulphur atom as at times {46} two-handed; and accordingly it was again to be assumed that such an atom who has engaged both its hands and is mindful of the phrase "hold fast to what you have"[136] will, following this phrase as a guideline for its actions, be wary (or be incapable) of grabbing at additional atoms.

But what if it does? As unbelievable as this appear to us, in a few instances it seems to happen. "This case is uncommon, gentlemen," a long-dead colleague used to say, "but it happens, and in fact it happens often."[137]

We inquire of the nearest attendant whether something like this can be seen in the aerarium. *Ma, sicuramente!*[138] is his answer. He shows us a balloon in which

---

[132] There are apparent allusions in this sentence which defeat my efforts to interpret. Is Kopp just having some deliberately mystifying fun with his readers here?

[133] No one is obliged to do more than he is able.

[134] There is no obligation to do the impossible.

[135] All three of these Latin expressions signify: he who has nothing, gives nothing.

[136] Revelations 3:11.

[137] I have not been able to trace this reference.

[138] Italian: Certainly!.

molecules are swimming where for each molecule one oxygen atom has approached two methyls (the compound as a whole answers to the call dimethyl ether; it is very volatile, and the demonstration is to be done at a very low temperature, let us say at almost at the freezing point of water); the two hands of the oxygen atom rest in the two hands of the two methyls. With a skill that would be entrusted to only one professional chemist now living[139] he extracts a certain volume, containing a certain number of molecules, from the balloon into a small specially constructed vessel, then adds an equal volume, hence an equal number of molecules, from the balloon that contains hydrogen chloride gas, of which each, as we know, consists of a one-handed chlorine and a one-handed hydrogen, that are associated together by a hand-bond. How remarkable: now the entire assemblage in the small vessel occupies a smaller space than that given by the sum of the two volumes combined together; one could conclude from this—or at least it has been concluded—that now fewer {47} molecules are present than the sum of the molecules of dimethyl ether and hydrogen chloride, and that could only be possible if for some of the oxygen atoms each of them that previously held two methyls also seized a chlorine and a hydrogen atom and (so to speak) exploited them to form a more complicated compound molecule.[140]

Seized; but with what, if each oxygen atom only has two hands and these are already held in other hands?[141] "With two supplementary hands," it is said. But that is really to say nothing at all. Where do they come from? Are they perhaps lent to the oxygen atom on this occasion, in addition to the two hands already possessed, merely honorarily? And then the oxygen atom is no longer two-handed but rather classed in the higher category of the true (not just as e.g., the thus-characterized) four-handers. Or is it a secret four-hander, who normally does not make use of this particular character but rather conducts itself like an ordinary two-hander? That would be something which we do not often see in the human world. But some believe that this sort of thing often happens in the molecular world, that the same atom can possess a larger number of hands, of which it normally only makes partial use, so that it has the appearance of really having the number of hands customarily used; but when it has already engaged these hands for the holding of others and then something is offered it that attracts its fancy, it then applies the others, or perhaps at first just some of them. This view has many adherents. It is not only rumored but some

---

[139] Surely Bunsen is meant here.

[140] The modern name for this substance, stable at low temperatures, is dimethyl oxonium chloride, an addition product of dimethyl ether and hydrogen chloride. A contraction occurs vis à vis the sum of the gaseous volumes of the two reactants, since the reaction reduces the total number of molecules by half.

[141] This sentence, and the paragraph that follows, raises the issue of the variability of valence, a subject much discussed in the 1860s, 1870, and 1880s. Kekulé, for one, attempted to maintain a doctrine of strictly constant valence, using such auxiliary hypotheses as "molecular compounds" (such as the substance treated in the immediately preceding paragraph), whose pieces he conceived were held in a molecular union not by valence bonds, but by attractive *physical* forces. Others, such as Wurtz, Williamson, and Frankland, explained the same compounds by suggesting that the valence of certain atoms could vary. For a good discussion, see Russell [24,171–223].

virtually assert publicly that an atom of a member of the chlorine family—chlorine, iodine, or bromine—is not at all one-handed in the strictest sense of the word, as {48} it mostly behaves, but when it wants to it can grasp with three or five or even seven hands. Many estimate the binding power of the sulphur atom as three times as large as we have hitherto assumed; according to this view, this atom usually only makes use of two hands, but possibly of four and even of six. Similar things are said and written about other elements.

To whomever such a view appears not just as a re-statement of the facts but as offering a satisfactory explanation of the facts should certainly make use of it and rejoice in its satisfactory character. He should not be swayed that the explanation of the relevant facts—only one of which was able to be mentioned here—can be attempted in other ways as well; for it is true, if not in all, certainly in most cases, the other explanations that one might otherwise attempt are equally vulnerable to objections, and not very satisfying. He should not be deterred by the difficulties which appear in the honest endeavor to pursue, develop, and understand this viewpoint in its consequences.

The first difficulty is this, that in recognizing a variation of the handedness by which a certain elementary atom functions it is probably not possible to arrive at certain conclusions about how many hands any given atom possesses. It is not possible for the reason that we can only make inferences regarding this capability of various atoms by comparing them among each other, and, if the possibility of a variation is stipulated at all, one may assume with some justice that the atoms of all elements possess this capability, and as a result we would no longer have any constant measure whatever for the comparative determination of the number of {49} hands of other atoms.[142] But to explain this in more detail would take us much too far afield.

The question can be discussed in a briefer way: assuming that view is correct, are or are not the hands that an atom always uses of the same kind as those that it only sometimes uses, and what does the atom do with the hands that it is not using?

One might imagine that an atom, let us say an oxygen atom, has four hands of two different types—metaphorically considered, two right and two left hands—and that it customarily seizes with the first kind, and does so more tightly with them. According to this idea, such an atom would hold more tightly two hands held with the right hands—in the case we just discussed, the hands of two methyls—and less tightly two others held with the left hands—the hand of a chlorine and of a hydrogen atom, each of which atoms once let loose, unite again to form a molecule of hydrogen chloride; and that would correspond to what actually happens. So if the view we are considering had to be adopted, I could more or less reconcile myself with this first idea. Better that than the alternative notion, according to

---

[142] Kopp's point is that to admit variability of valence in principle suggests opening the door to a potentially indefinite number of possible structures, even for otherwise simple molecules. Molecular chemical theory would then become (in the view of some) both unwieldy and uncertain. This is undoubtedly the motivation for defenders of constant valence to maintain that doctrine, even in the face of numerous apparent exceptions to the valence rules.

which such an atom always allows two claws to be seen and used, keeping two others hidden in reserve, treacherously bringing them out and making use of them only when it wishes. Whereby it might remain an open question whether the following conception would be allowed: that such an atom really has only two hands, but on each hand, in addition to the true and stronger thumb, it has another weaker one situated opposite to each of the other fingers. With atoms for which— as for the oxygen and sulphur atom—the handedness is assumed to vary always by even numbers, the first conception can be more easily carried out than for atoms {50} that are said to grasp always with odd numbers of hands, such as e.g. the atom of phosphorus which is considered as behaving sometimes three-handedly and sometimes five-handedly; then the matter would not turn out so symmetrically. And even more awkward, when it concerns an atom, like nitrogen, that in various compounds makes itself out as if sometimes it has an odd number of hands, three or five, sometimes an even number of hands; everything stops unless one makes the assumption that the atom sometimes decides to cover the thumb of one hand.

But apart from that, what does an atom that has more hands than it uses for gripping and holding do with the free ones? Earlier (p. {10}) we noted that atoms have a very powerful urge to grip; how can this be reconciled with the present case? If the number of free hands is even, it's probably again simplest that each hand grips and holds another; if one should be left over, it would do what it can to behave according to the just-stated assumption.

Some are of the view that atoms are not endowed with any fixed number of hands, but that each atom grips with as many hands which it needs for seizing as are good for it, according to the nature of those that are offered to and accepted by it. This view seems somewhat doubtful, and not just because it reminds one somewhat of the doctrine of Fourier, according to which inclinations and capabilities should be proportional to one another,[143] but also for other reasons. The ability of an atom to grip with a certain number of hands surely ought to be considered as inherent to the respective atom itself and determined by its nature, even though it should be admitted that the application {51} of this capability, namely with how many of its hands the same atom can grip in a given case, depends on the nature of the atoms offered to it, as well as on other circumstances—temperature and such. For without this the entire doctrine of the handedness of atoms probably would not be able to make a claim for its existence.[144]

There are again some, but fortunately not many, who think that this doctrine at the present time has to be sure a good deal of hand, but not very much foot[145] to it. The doctrine, they say, debuted with simple assumptions, and the conclusions from it have been strictly drawn, definite, and, as far as connecting and predicting facts

---

[143] I.e., that there should be a rational division of work in order always to provide the most appropriate jobs to the most appropriate people, as advocated by the utopian socialist Charles Fourier (1772–1837).

[144] See note 142.

[145] I.e., long-term viability.

are concerned, quite important. But then, they continue, the facts that do not agree with it have been better noticed, and newly found facts that are equally discordant have been added, and through experiments to adapt the doctrine to all this, it has not exactly been perfected. We have, it is said, something better under the previously (p. {21}) described viewpoint according to which we judge theoretical doctrines; however, too, since …[146]

But for the last little while you have not contributed a single word to the conversation, and now you only ask what I think about this doctrine, and especially about the facts that pose so many difficulties? About the middle of the 1820s a popular ballad was sung to the sound of a hurdy-gurdy, illustrated with a picture-board dripping with blood, which popularized the atrocities of the crafty and murderous Mausche Nudel,[147] and more than one verse, after depicting the unbelievable crimes that were committed, concluded with the words, "How this happened was wondrous; Patience! All will be clear with time." {52} Probably to him who has learned to wait from an early age. But now we want to use our time more economically, so as far as possible we will keep from digressing and will look further at what the aerarium offers.

[The Story of Iodine and Hydrogen]

The products of dissimilar atoms or atomic groups that are formed by handholds do not last forever. "This too shall pass" is sung not only in the human world. A dissolution of the union can be caused by the entry into it of other atoms or atomic groups that previously were members of other molecules. We cannot in any fitting way describe all that happens here; we satisfy ourselves with the reminder that elective affinities are exerted there that cannot at all be compared to those described in a well-known novel which a critic of that time called a chemical decomposition of sin.[148] But quite often a modest rise in temperature suffices to separate that which till then was united, whereby each part thus separated looks around to see where he can once more find lodging through handholds. We shall consider only one case of this, but it is an instructive and in many respects not uninteresting case.

Like so many elements that later rose in chemistry to an esteemed or even significant position, iodine too can be considered a foundling child. For the mother liquor of a certain salt in which a brave saltpeter-boiler found it in Paris in 1811

---

[146] Once again we have a deliberate self-parody, pointing to Kopp's tendency to obfuscate and delay giving a forthright judgment. The next sentence suggests Bunsen politely interrupting, asking Kopp please to get to the point.

[147] *Sic*, for Mauschel Nudel, a.k.a. Isaac Moses, a.k.a. Moyses Hoedt, a notorious Frankfurt criminal at the beginning of the nineteenth century. In the middle of the 1820s Kopp would have been a boy of seven or eight years of age.

[148] "Eine chemische Zerlegung der Sünde." The critic was Wilhelm Grimm, describing Goethe's 1809 novel *Die Wahlverwandtschaften* (elective affinities), in which a chemical metaphor is overlaid upon a complicated love story.

was not at all assumed to stand in a closer relation to it.[149] A few years later various distinguished men[150] took its part and announced it to the chemical world, in which it quickly learned how, so to speak, to make itself indispensable. {53} The name given to this element, iodine, is from το ίov, violet, derived in a similar but not exactly the same way as rosine from *rosa*, pink. But already very early on, the young thing was called iod or jod; the famous [Joseph Louis] Gay-Lussac, who was fulfilling his role as godfather in inventing an appropriate name, called it that, and this name was entered into the official chemical birth registry of the city of Paris. I believe everyone uses the shorter designation, except those who make use of the English language.[151] Germans usually do this too, who by the way could have named the new element much more simply "Veilchen," since this given name exists in Germany (the human Violets all have dark eyes). The similarity between this element and the plant is, to be sure, not great. Of which C. Schreiber spoke in his poem "The Language of the Flowers":

> Do you know the violet, the blossom of May?
> Modesty gave it the costliest price,
> Seen only by the eye of humility,
> It blooms out of sight, but smells so lovely.[152]

None of this applies to the element in question. The colour of iodine vapor is, to be sure, violet, but there the similarity ends; all else that redounds to the fame of the violet plant is not shared by iodine, and as for the smell—we should best not speak of that.

When iodine was some years old, its ancestry was revealed. This happened not in what could be called the coarser way, according to which such identifications happen in the human world. In these instances very often birthmarks are preferable for evidence. One of the best known among the relevant {54} cases is that of a

---

[149] In 1811, Bernard Courtois (1777–1838), a Paris saltpeter manufacturer, discovered iodine in the mother-liquor prepared from the ashes of saltwater plants, but he does not seem to have regarded the substance as a new element [25].

[150] These were Humphry Davy, Nicolas Clément, Charles-Bernard Desormes, and Gay-Lussac. Gay-Lussac had named the new substance "iode" (from the Greek for violet), carefully chosen so as not to suggest anything about the composition of the substance. In December 1813 Davy and Gay-Lussac each concluded, probably independently of each other, that iodine was elemental. Unluckily, each suspected the other had stolen his idea and sought to publish first. A good summary of this affair is in Partington [6 4:85–90].

[151] Of English, French, Spanish, Italian, and German, only in English is a suffix added to the roots chlor-, iod-, brom-, and fluor-, to make chlorine, iodine, bromine, and fluorine. This pattern was begun in 1810 by Humphry Davy, when he first suggested naming "oxymuriatic gas" by the neologism "chlorine," whereas Gay-Lussac, adopting both Davy's view of its elementarity and the neologism, preferred "chlore" in French. When Courtois's discovery was promulgated, Gay-Lussac chose the parallel French "iode," whereas Davy in 1814 (opining that "ione" in English "would lead to confusion") preferred "iodine."

[152] "Kennst Du das Veilchen, die Blüthe des Mai's? / Sittsamkeit gab ihm den köstlichen Preis! / Nur von dem Auge der Demuth gesehn, / Blüht es verborgen, doch duftet es schön." From Christian Schreiber (1781–1857), "Die Sprache der Blumen" (1805).

certain *Preciosa*. We know (old habit compels us to go to the original sources and not rely on popularized versions), from the first of the *Novelas exemplares* of Cervantes, how the *Corregidora* or female city magistrate of Murcia established the identity of the girl who was associating with gypsies as that of her lost daughter, that there could be found on her breast a rubrication, *una señal pequeña, à modo de lunar blãco, cõ que avia nacido*, and how this mark was verified, elucidated, in addition to the text of the Antwerps edition (itself called priceless by Brunet) of the cited work of 1743, also by a clear copper engraving.[153] Which does not exclude the possibility that other distinguishing characteristics of the same kind brought to the world could be provided; just to stay with the cited case, the mother also found by the elongated shape of the foot the confirmation she sought: *que los dos dedos ultimos del pie derecho se travavã el uno cõ el otro per medio cõ un poquito de carne.*[154] And by the way, in the absence of such congenital marks there are others that are not to be despised when sufficiently individualized, such as more-or-less clear scars at certain locations; in one case told of in a poem that begins "In Myrtill's ruined hut the lamp was still gleaming," and whose title and author I have unfortunately forgotten, where the mother finally recognizes her Walter: "Let me see the pockmarks! Yes, it's you, my lost son!"[155]

But it was not by such external marks that it was recognized to what family iodine belonged. Rather, as its chemical character was better understood, it was no longer at all doubtful that it was closely related to the already-known chlorine. And in any case {55} one external characteristic was similar: iodine, like chlorine, is one-handed.[156]

One cannot say that iodine grew up in the free state in isolation. Rather, in this state it was and is distinctly accustomed to duality, and in it under normal circumstances it is seen hand in hand with an identical sister: two iodine atoms unite to form an iodine molecule. This union is of course dissolved, and one sister must leave the other or at least withdraw her hand, whenever one of them—or, as generally happens, both simultaneously—on the basis of so-called affinity enters into other combinations. These combinations, it must be admitted, are very numerous. We will examine only one, that with hydrogen.

All who can more or less judge what simple atoms think and feel agree that neither free iodine nor free hydrogen—to whose molecules as we say two atoms

---

[153] Miguel de Cervantes, *Novelas exemplares, en esta nueva imprecion* (Antwerp: Bousquet, 1743), the first of whose short stories was "La Gitanilla" (the Gypsy Girl, 1613); Jacques Charles Brunet (1780–1867) was a famous French bibliographer. The Spanish phrase means "a small mark, in the form of a white spot with which she had been born."

[154] "… that the two last toes of the right foot were connected together in the middle by a little bit of flesh."

[155] "Walter der verlorene Sohn" (ca. 1790), by Johann Friedrich Schlotterbeck (1765–1840). Considering its outrageously digressive character, this entire paragraph can only be viewed as a deliberate self-parody.

[156] The chemistry of iodine was well studied by Gay-Lussac in 1814. Its monovalent character was recognized upon the introduction of ideas on valence in the 1850s.

are each joined—has any inclination to unite together. Apart from the fact that both are completely different from the other in their intrinsic tendencies, they also exhibit a certain external disproportion: the iodine atom is 127 times heavier than the hydrogen atom to which it is to be given. By means of artificial warming both can be made somewhat more compliant, and they unite apparently voluntarily, but somewhat sluggishly, and without yielding completely to each other—as one only partly expresses it. But it is always ominous when a combination proceeds with distinct cooling, and external heat must be supplied to make it happen at all; separation happens all too easily, and {56} in many cases explosively. At lower temperatures iodine and hydrogen are more inclined to combine when Mr. Platinum-Sponge is present and active; he acts as shadkhen[157] and exhorts both sides until the doubts are resolved. But the union happens especially when the paired iodine sisters meet a compound that hydrogen has entered with something else[158]; hydrogen often seems to be an unsatisfied husband (in what marriage would he not complain?), longs for change, and makes advances to iodine, which persuades her to try it with him.

So: the association comes to pass on the one or the other side—we won't waste our valuable time to remind ourselves of each alternative. The compound is quite colourless, acid, and airy, but under normal circumstances it behaves moderately well. Not when heated. Let's look into this balloon, whose temperature is indicated by a thermometer within; it is the home of the couple we speak of: iodine hydride, or as it used to be written, hydriodine (which should not be confused with the little Hydriot, to which in my youth Wilhelm Müller directed one of his very beautiful Greek songs[159]); an attendant is just lighting the fire underneath it. The atoms united to hydrogen iodide gas molecules are walking (not very quickly; the hydrogen iodide gas molecules move eight times as slowly as hydrogen molecules under similar conditions) in a straight line, and each atom contained in such a molecule, the iodine as well as the hydrogen atom, has also its own independent motion. It's getting warmer; first we see only that the molecules hurry somewhat {57} more than they did at a lower temperature. But by looking more closely we cannot miss seeing that the unity of those still strolling hand in hand appears to be somewhat disturbed. Are the intramolecular movements of each atom not becoming more independent? Does not each one jerk spasmodically as if it were saying, "I can't take it any more?"

Listen: I fear this will end as it does during peasant festivals, with acts of violence. The iodine appears truly to desire restitution *in integram*,[160] and the

---

[157] Hebrew and Yiddish word for matchmaker. Platinum sponge ("Herr Platinschwamm") catalyzes the reaction between hydrogen and iodine to form hydrogen iodide.

[158] A more convenient way to make hydrogen iodide is to react iodine with hydrogen sulfide.

[159] "Der kleine Hydriot," by Wilhelm Müller (1794–1827). A Hydriot is a resident of Hydra, an island in the Aegean.

[160] Restitution to the original state.

hydrogen appears to have had the sort of experience in this combination that
makes it seem that it would not be too great a misfortune if he would have cause
for making a change. There!—the thermometer indicates not quite 200°—an
iodine tears her hand from the hitherto associated hydrogen, and other molecules
follow the example of this one. That truly seems to qualify for malicious aban-
donment (*desertio malitiosa*)! For the two atoms who hitherto sought their way
together through the molecular world do not hurry separately and voluntarily on
their crooked path in such a way that one could be assured of their reuniting;
before the separation, the one does not speak to the other as do the two friends
while skating, according to Klopstock:

> If you turn left, I will
> Circle halfway to the right;
> Take the turn as you see me take it:
> Okay! Now fly right past me![161]

No! Each of the iodines separating from her former companion takes by the hand a
similar sister who is becoming free at the same moment, and the two hurry
stubbornly on in a straight line, turning blue in malice. Hydrogen behaves
essentially the same, though without turning colour. Look at that iodine just
becoming free: {58} isn't she making a face to her hydrogen, as if she were calling
out to him the words, from the song probably never sung to the harp by the higher
host, but probably sung in 1836 at the Mannheim May-Market:

> Just go away, you fickle one!
> Think no more thereon?[162]

The temperature is rising ever higher, and with it the number of hydrogen iodide
molecules that have ceased to exist because the iodine and hydrogen hitherto
contained therein have separated from one another. The number of these molecules
to which dissimilar atoms are still held in association by so-called affinity becomes
ever smaller; hydrogen gas and iodine gas molecules, to which two similar atoms
are bound by so-called friendship, stream forth in ever greater numbers. We should
not be deceived: what is going on here is worse than an alteration of government,
or revolution; it is dissociation!

   We are horrified to consider how many compounds can undergo dissociation
at a moderate temperature (but can one use the word "moderate" without protest
for such an immoderate, disorderly process?). And this not only for compounds,
like those just described, for which the circumstance that they can be created
directly only in cold conditions suggests that they will not be especially

---

[161] "Zur Linken wende du dich, ich will / Zu der Rechten hin halbkreisend mich drehn; / Nimm
den Schwung, wie du mich ihn nehmen siehst: / Also! Nun fleug schnell mir vorbey!" From
"Der Eislauf" (1764), by Friedrich Gottlob Klopstock (1724–1803).

[162] "Geh' nur hin, du Flattersinn! / Denke nicht mehr dran, / Wenn ich einstmals Wittwer bin /
Frage wieder an." From an old German folk song. The May-Market in Mannheim is a large
regional fair, held annually since the early seventeenth century.

permanent, and will not offer much resistance to destructive effects; but also for compounds in whose components such energy and warm feelings inhere that they release a portion thereof when they are united. But we have seen enough of such processes in this one case, and tarry no longer to further consider such instances. {59}

But our thoughts still cling to the hydrogen iodide molecule that our eye first followed, or rather to the atoms of which it was composed. Now they are separated from each other and float in the space within the balloon, whose temperature, as far as the thermometer lets us judge, has become stationary at 400°, each with a similar atom, united to a molecule, among uncounted other such molecules and still-undecomposed hydrogen iodide molecules. Will they find each other again? Will it be fulfilled for them as happens in the human world what is told of in many a poorer or better novel? Will this iodine after the separation gradually learn to understand what she really had in this, her hydrogen, even though he was no man of great substance in the molecular world, and all in all was a rather light character? Will he come to similar reflections regarding she who is no longer there? And most importantly: will the estranged come back together, perhaps—which would be the loveliest because it is most improbable—feel interest for another, and without recognizing each other reunite? We must recognize the possibility at least of the latter conclusion of the story: that it could happen; but also the improbability that it will.

[Friendship under Extreme Heat]

In the molecular world, it is very carefully prescribed according to what numerical proportions molecules of an original compound—in this case, hydrogen iodide gas molecules—may and should come together in the same space or portion of space, and similarly regarding the new molecules created by dissociation—in this case, hydrogen and iodine gas molecules. This proportion changes with the temperature, as we can already see from the preceding; at higher temperatures more {60} hydrogen iodide gas molecules are decomposed, and accordingly more hydrogen and iodine gas molecules are present, than at lower temperatures. The total number of molecules inside a space of a certain size also exerts an influence on this proportion, and whether the space is pressed more or less full, or (as we say for short) what the pressure is. But we do not now have time to treat the influence of the latter; we will however spend a few minutes in considering the effect exerted by temperature. We know that the given rules pertaining to this proportion, the so-called equilibrium proportion, are strictly observed, and that in every part of the space, as it were, a constable is posted, who in the event of a violation takes down the names of all the molecules (or atoms associated therewith) present in his division, so that the guilty can be identified and brought to justice.[163]

---

[163] Kopp's description of dynamic reaction equilibria in this and the next paragraph must be seen in the context of developing understanding of the subject in 1882. This was still early in the history of the subject, but Kopp could use the work of Alexander Williamson, Ludwig Wilhelmy, Marcellin Berthelot & Léon Péan de Saint-Gilles, Augustus George Vernon-Harcourt,

Just as for every space inhabited by people, so also for every space filled with gas molecules the temperature is not the same at every spot; rather, some spots are a little warmer and some a little cooler; what is observed to be the temperature of such a space is that of only one spot, or at best the average of all the various parts of the space. The equilibrium proportion may be established for a moment at a cooler spot; we imagine what was present in this spot, or only a portion of it, when in the next moment it becomes a hotter spot: to re-establish the equilibrium proportion for the temperature of the latter spot, in the most summary process of the divorce proceeding an additional number of yet-undecomposed hydrogen iodide gas molecules must be introduced, a decision reached, and a verdict for disposition brought, in order that the number of the latter molecules {61} versus the number of hydrogen gas and iodine gas molecules shall then be present, all according to the regulations. And vice versa: in the transition from a hotter to a cooler spot a number of hydrogen gas (or iodine gas) atoms, which till now were united in a friendly molecular union with a similar atom, must immediately enter into combination with a dissimilar atom on the basis of so-called affinity, for a certain number of new hydrogen iodide molecules must be created to establish the equilibrium proportion corresponding to the lower temperature. And it must happen quite quickly; all the best intentions don't matter, only results.[164] To a dilatory molecule whose turn it is to be transformed the threat is to take down its name, or even be sentenced to detention, which is the hardest punishment for gas molecules, who love to stay in motion.

Thus the haste with which the separations of similar and the unions of dissimilar atoms occur. Hydrogen exhibits a decisiveness of will and action which we have not seen in many other cases, qualities that he shows to the very first iodine who comes (with her sister) across his path. The iodine, too, is not the least bit shy; to hydrogen all her body language says yes, and gives her hand, and she acts as if she had never been in any other than the free state. There are no further preparations nor ceremonies preceding the combination. She has no need to arrange a dowry, and the bridal wreath is neither bound nor sung. It is not our place to question whether this is appropriate; it would be up to those who are directly involved. But it certainly would be lovely if this custom were to be reinstated in the molecular world, and it would be of great interest to know whether {62} the green wreath is bound with violet silk there as well; a delicate attentiveness would lie in the choice of just this combination of colours—which in general one might perhaps consider not the most fortunate—for the particular case of iodine.

(Footnote 163 continued)

C. M. Guldberg & Peter Waage, and Leopold Pfaundler. Within just a handful of years after the publication of *Molecular-Welt*, the subject was enormously advanced by van't Hoff, Arrhenius, and Ostwald. That said, Kopp's qualitative descriptions are remarkably consistent with later understanding of the influence of temperature on competing equilibrium reactions.

[164] "Maulspitzen hilft da Nichts, gepfiffen muss sein." German saying; literally: it doesn't do any good just to purse one's lips, one must whistle.

"Lavender, myrtle, and thyme," you say?—my, what a naturalist you are![165] But if the bridal wreath is sung at all in the molecular world, then surely it is done not in the particular language of any one country, but in the universal language, Latin, and what you are thinking of will probably be uttered in the approved translation:

> *In horto frutex Veneris*
> *Cum thymi flore crescit.*
> *Quod tempus terit juvenis?*
> *Spes vivida languescit.*
> > *Belle virens sertulum,*
> > *Sertulum canamus!*[166]

as far as I remember it; but it has been a long time since I heard it. But to come back to the main point, what is certain is that at constant temperature, in the various warm parts of the space just as many hydrogen iodide molecules are formed anew in each time interval as are decomposed, and when the temperature is dropping, more hydrogen iodide molecules are formed than are decomposed. To be sure, it is also possible that a hydrogen atom comes together and joins with the same iodine atom with which it previously had been united; but in the hurly-burly in which all this takes place there is very little probability of this.

Should similar atoms united by friendship into a molecule hold together more securely under circumstances of high temperature than do dissimilar atoms brought together into such a compound by so-called affinity? {**63**} Should it be said of this, as an old proverb expresses so harshly:

> Friendship is like an old walking-stick on a journey;
> Does the stick enjoy the walk?[167]

In fact this has been believed for a long time, at least as a general rule. But observations of recent years have—might I say, corrected this impression? No!—they have further enriched us with sad experience.

We have mentioned chlorine several times, especially noting that two chlorine atoms unite to form a molecule of chlorine gas by joining hands. In this union, in the so-called free state the chlorine sisters behave quite freely (they seize hold of everything) and in a rather emancipated fashion; if they took it so far that they would even join in on student songs, they could even take, even if not entirely appropriate to them, as an expression of their feelings, the beautiful song, according to the solemn melody of a hymn that E.P. and I once heard sung on a

---

[165] "Wir winden dir den Jungfernkranz / Mit Veilchenblauer Seide / ... / Lavendel, Myrt' und Thymian / Das wächst in meinem Garten / ... / Schöner grüner Jungfernkranz / Mit Veilchenblauer Seide." (We bind the bridal wreath for you / With violet silk / ... / Lavender, myrtle and thyme / Grow in my garden / ... / Lovely green bridal wreath / With violet silk.) From Carl Maria von Weber's opera *Der Freischütz* (1821).

[166] I have not been able to identify this Latin verse.

[167] "Freundschaft ist ein Knotenstock auf Reisen, / Lieb' ein Stäbchen zum Spazierengehn?" From Adelbert von Chamisso (1781–1838), "Das Lied von der Freundschaft."

Sunday morning in the Franciscan monastery of Cimiez near Nice,[168] which put all present in a contemplative mood:

> Where brothers frankly and freely
> Bound by brotherly sentiments
> Of the rosy bonds of friendship
> Harmoniously hold hands fast[169];

for these atoms remain undivided even when strongly heated; less so when other atoms come between them.

"Benevolent is the power of fire, when it is mastered and guarded by man,"[170] these molecules could also say, with some particular emphasis on the word "mastered." When a fire is kindled below them, they feel at first much better than before. With the rise in temperature they feel themselves rising ever more {64} to the completely ideal gaseous state,[171] and in this feeling they cover ground all the more nimbly. The heat puts wings to their hurrying feet; if the acceleration of tempo continues, we can expect that the motion of the molecules will be half again as fast at a little more than 340°, and twice as fast at about 820°, as at 0° (this increase in speed with rising temperature happens equally with the molecules of all gases).[172] But the latter temperature, 820°, is very high; in this state (the heat is already such that the glass of the balloon will surely soon begin to soften)[173] do the atoms bound together by friendship into molecules still hold together? The attendant is changing the vessels in which until now the airy company was held; part of them are conducted into a small container constructed of the best porcelain, in order to test the stability of the molecules in the fiery furnace. No matter; our mind's eye can also see through the porcelain.

We do not yet know how furnaces were constructed in antiquity.[174] Even in the first century A.D. the descriptions of such useful devices are sketchy; according to Pliny, various kinds of furnaces were already being constructed for metallurgical purposes, and this impressive polymath distinguished the sides (*latera*), the interior (*camera*) and the mouth (*os*) of smelting furnaces, from which we cannot

---

[168] The Franciscan monastery still stands, though Cimiez is now a neighborhood of Nice. "E.P." is not identified; possibly Eugène Péligot (1811–1890)?

[169] "Wo sich Brüder, fest umwunden, / Von der Freundschaft Rosenband / Und durch Brudersinn verbunden, / Traulich reichen Hand in Hand." From "Wo zur frohen Feierstunde," an old German folk song popular among university students.

[170] "Wohlthätig is des Feuers Macht, wenn sie der Mensch bezähmt bewacht." From Friedrich Schiller (1759–1805), "Das Lied von der Glocke" (1799).

[171] Divergence of real gases from ideal gas behavior is greatest at low temperature, least at high.

[172] Since mean molecular kinetic energy, which is proportional to mass times velocity squared, increases in direct proportion to absolute temperature, molecular velocity is proportional to the square root of temperature. These three Celsius temperatures correspond to 273°, 613°, and 1093° degrees on the absolute scale, whose square roots are as $1:1\frac{1}{2}:2$.

[173] Ordinary soda-lime glass begins to soften around 700° C., borosilicate glass around 820°.

[174] Nor do we know much more about this subject today.

however learn the details of how such a furnace was built.[175] We know much more about the construction of furnaces in later eras, so much that we cannot go into the subject here. But in this respect the laurels go to what has been invented in the modern era. What is accomplished in convenient form by a Perrot gas furnace with a Wiesneg [sic] muffle[176] {65} appears to satisfy all reasonable requirements for a truly infernal heat; the most intense yellow heat, reached at a temperature of between 1500° and 1600°, is no trivial matter.

Into such a furnace is carried the little porcelain prison in which there are a not inconsiderable number of chlorine sister couples; the heat rises fast. Even at the start of the heating we note the increasing speed with which each molecule hurries onward, also that the movements belonging to each of the atoms forming the molecules has become livelier; but nothing suggests to us that the unity is endangered, that a separation is about to occur. Until almost 600°, until it begins to glow a dull red, the two atoms joined hand in hand together into each molecule remain united. But when the red colour becomes a bit brighter some of them appear to become uneasy. Consider this one molecule just brushing by us now: how each of the atoms looks at the other and how it behaves while moving, is that not like the whole pantomime performance of the feeling expressed by the first few words of the poignant song which one could hear years ago, and probably still can today, in the Odenwald:

> Don't think ill of me that I am avoiding you,
> And release me from every obligation?[177]

And they do now actually separate from each other; but each chlorine seems, having had to separate from its sister, to come closer than we really could expect after the earlier observed behavior of the hydrogen atoms that associated on the basis of so-called affinity with chlorine atoms to form molecules. And many other molecular comrades also are now parting, in proportion as the temperature rises higher and higher, more and more follow the same example; at the very highest degree of heat {66} that this particular furnace can reach there are only very few of the originally present molecules still unchanged, and we cannot doubt that at still higher temperatures even these molecules would not be able to escape decomposition into its atoms. This is once more the detestable dissociation that brings this about, here tearing sister from sister, there as in other cases destroying molecular stability. There you can see its victims, the one-handed chlorine atoms, wandering about, no longer finding a sisterly hand in which to place its own. And what can each do with its newly freed hand? Perhaps beckon to her chum, now gone? Forsooth, a small consolation! And can a substitute for the pain of separation be found in the prospect that it probably will not remain so hot for long, and

---

[175] From Pliny the Elder's *Historia naturalis*, ca. 79 C.E.

[176] These innovations were introduced by Perrot and V. Wiesnegg, Paris instrument makers, in the 1870s.

[177] "Verdenk' mir's nicht, dass ich Dich meide, / Und sprich mich frei von jeder Pflicht." From an old German folk song. The Odenwald is a range of low well-wooded mountains southeast of Darmstadt and northeast of Heidelberg, mostly in Hesse.

when cooling begins, a larger number of chlorine molecules each formed by two joined atoms (corresponding to the lower temperature) will be present, so that for a currently solitary chlorine atom there is hope to come back together in a renewed union with what it had hitherto been associated? But it is not only, as it was stated so well in the last decade of the last century in a family album dear to me, that "hope is a long rope on which many have pulled themselves into foolishness."[178] Rather, when such happens, there is the greatest probability that a chlorine will join with some other to form a molecular union, than with exactly the one with whom she had wandered hand in hand before the separation.

Those elements belonging to the same tribe, whose gas molecules also consist at low temperatures of two atoms joined together, behave in strong heat very similarly to chlorine. {67} Thus iodine, whose molecules begin to lose their mortar (or better, decompose into atoms) at an even somewhat lower temperature than those of chlorine. So also with another element of the same tribe, of which I have deliberately avoided speaking in more detail until now; it has justly been given the name bromine, for indeed it is a very evil-smelling element.[179] But even when the gas molecules of elements that are not member of this family are heated to great excess, the atoms composing them contemplate whether it might not be better to separate from the molecular union, and under the increasingly urgent conditions of temperature allow the secession to become fact. For instance, this is what happens for the molecule of phosphorus vapor, which is stable at a relatively low temperature and consists of four atoms; at a very high temperature this four-leaf-clover molecule begins to lose its leaves one by one.

But not all gas molecules consisting of several atoms exhibit this property of ultimately decomposing when the temperature is continually raised. For example, the molecule of ordinary oxygen gas, which is composed of two oxygen atoms, remains untouched, its atoms not parting from one another, even at the highest temperatures at which its stability has been studied. It is comforting to find also in the molecular world something of which one can say: when everything else becomes false, we will still be true. It scarcely need be emphasized that this sort of alteration is not possible for the mercury molecule, which consists of but a single atom. Considering the entire disposition of the mercury atom/molecule, just as unnecessary as is the warning that a constable {68} might make to the same, "Don't gang up together," equally impossible would it be for it to obey the command, "Disperse."

---

[178] "Hoffnung ist ein langes Seil, an dem sich schon Mancher zum Narren gezogen hat." Similar to the German proverb, "Hoffnung ist ein langes Seil, an dem sich viele zum Tode ziehen" (hope is a long rope on which many pull themselves to [their] death).

[179] In 1825, Antoine-Jérôme Balard (1802–1876) discovered bromine, a new liquid element, from seaweed ash; Gay-Lussac named it "brome" (German "Brom"), from the Greek word for "stench," after its foul smell. The English immediately added the –ine suffix, parallel to chlorine and iodine. Fluorine had already been hypothesized, though not yet actually isolated in elemental form. It now became fully clear that these four elements formed a natural family, like the metals; Berzelius named them "halogens," which means "salt-forming" elements.

[Molecular States]

Until this point, we have considered gas molecules as single entities: how individuals are composed, how they move, how they part. We have until now been concerned with what could be called matters of civil law in their relationships; but we do not want to leave this part of the aerarium without also thinking about that which concerns the common existence of many such individuals within the same space, namely arrangements that pertain to the public laws that govern a molecular state.[180]

First and foremost we must note that the constitution[181] of such a state is democratic on the broadest foundation. As a matter of principle, no atom enjoys a preferred place above any other; nothing is known of rights of birth or of class, such is not demanded or granted even on grounds of convention or of courtesy. In every state, even those with a mixed population—where molecules of unequal weight are assembled together—an equal *vis viva*[182] is assured for all molecules. The organism of the state functions without any special official authorities. Each molecule administers the law himself; when earlier I spoke of constables posted to keep peace in various parts of a molecular state, this was on the one hand just a turn of phrase, on the other hand intended in the sense that whenever there is a disturbance of the public order, every molecule in the neighborhood is as justified as it is obligated immediately to step in and take care to apply the necessary remedy.[183]

The constitution of a molecular state is essentially distinct {69} from other constitutions, especially from putatively better constitutions of human states, in that in it there is precious little about fundamental rights, but much more about

---

[180] Kopp's nonce term is "Molecülen-Staat," meaning a polity or state (Staat) consisting of a collection of molecules. Of course, the English word "state" can mean not only "polity" but also "state of matter." This circumstance creates a double entendre in English that is not present in the German, for "Staat" can only refer to a political organization. Germans use a different word for "state of matter," namely "Zustand."

[181] Kopp's word here is "Verfassung," which means "constitution" in the sense of the fundamental law of a polity (or, metaphorically, one's general condition of health). "Verfassung" would not normally be used to designate the structure or internal composition of a body; that meaning is conveyed by a different German word, namely "Konstitution" (which, perversely, can also mean Verfassung, the legal framework of a country). In English, "constitution" can mean either internal structure, or basic law. Consequently, once more we have better puns here in English than in Kopp's German.

[182] I have translated Kopp's term "lebendige Kraft" by "vis viva," for the latter Latin expression has the same technical as well as literal meaning as the former German term. The technical meaning is kinetic energy (proportional to mass times the square of velocity). The literal meaning is "living force," so it is an intentional double entendre.

[183] In 1871 Kolbe and Erlenmeyer had used a similar metaphor, but applied it to organic molecules in condensed phases, rather than to gas molecules. Kolbe [12, 128–29] analogized the atoms in these molecules to "a well-organized constitutional state, with one sovereign and a number of subordinate members ..." Erlenmeyer disagreed [26, 32–33]; the atoms in such molecules "find themselves in a condition of mutual dependency ... every elementary atom has a seat and a vote whenever the chemical fate of the state is to be decided ..."

fundamental obligations of molecules. Among the fundamental duties of molecules is, above all, that each move in a rectilinear course with a velocity appropriate to its weight and to the temperature, and that each exert its constitutionally
determined effect at the border of the state, when striking against the wall of the
space. This effect of each molecule in collision with—and exerted as pressure
on—the wall is partly determined by the violence of the blow, which in turn is
determined by the weight as well as the speed of the molecule, proportional to the
product of mass times velocity; but it is also determined by how often molecules
strike the wall in a given time, and the number that expresses this is again clearly
proportional to the velocity of the molecule: the effect of each molecule in
collision with—and exerted as pressure on—the wall, taking everything into
account, is proportional to the product of the mass of the molecule and the square
of its velocity, which is its so-called *vis viva*. It is therefore clear that for one and
the same volume of gas, the magnitude of the pressure against the wall must be
proportional to the number of gas molecules contained within the space, and must
increase with increasing temperature if the number of molecules occupying the
space is constant, assuming that the velocity of the molecules increases with
temperature in the manner we spoke of earlier.[184]

As regards molecules of one or another type, according to international
agreement the different molecular states are obligated, under similar external
circumstances {70} (i.e., at similar temperature and pressure), to allow one and the
same space to be occupied by an equal number of molecules.[185] Assuming that the
temperature remains constant, if any particular number of molecules occupies a
certain space at a certain external pressure, keeping a certain average distance
apart in their motions, and the pressure changes, the molecules must accordingly
alter their average distance from each other, so as to make the volume they occupy
inversely proportional to the pressure. Assuming now that the pressure remains
constant, if for a certain volume of one gas and for an equal volume containing an
equal number of molecules of another gas, the originally constant temperature
experiences an equal alteration, the molecules of each gas must alter their average
distance from one another so that the volume of the one and of the other are also
equal at the new temperature.

These fundamental obligations of gas molecules, like the fundamental rights of
the members of a human state, are established under the assumption of an ideal
state, which molecular states can more or less approach, without in reality being
able to achieve it in total rigor.[186] To the credit of the members of molecular states,
it should however be truthfully reported that they overwhelmingly correspond
closer to the presumed state and behave far more correctly than do the members of

---

[184] On p. {64}.

[185] This is the fundamental hypothesis of Avogadro, proposed in 1811, but not generally
accepted as a scientific principle until ca. 1860.

[186] Again, in this sentence we thrice use the English word "state," but in Kopp's German the
first and third occurrences are "Staat," the second "Zustand."

human states, and do not so easily provide occasion for the suspension of constitutional measures by a declaration of a state of siege or even the proclamation of military law. The citizens of human states too often lack the desirable zeal to fulfill {71} not only the assurance of their fundamental rights, but also the performance of their fundamental duties. The citizens of the molecular state deviate, when they do not behave in the ideally correct way, according to the opposite side of the principle of strict observation of the prescriptions of the constitution; in their zeal to do their duty, they nearly always go somewhat too far, and consequently mistake serious matters for trivialities; one might even be tempted to admonish them with the words which the professor (also superintendent and inspector of schools) P. in G.[187] gave to the teacher who, during a visit by his superior, after accompanying him to the stairway, wanted so quickly to escape his eyes that he fell rather than walked down the stairs: zeal is praiseworthy, but moderation is to be desired.

For instance, when the external pressure is increased and therefore according to the constitution it is decreed that the molecules must approach each other more closely (or occupy a smaller space), the molecules of gases generally do more in this direction than they are obligated to do, they occupy a smaller space; that is, the diminution of space is greater than what corresponds to the increase of pressure. The molecules of most gases exhibit this often quite regrettable hyper-loyal behavior (even though it is based on a noble motive) ever more perceptibly the more the increase of pressure is continued.[188] Only the molecules of hydrogen gas exhibit less good will to obey the constitutionally-prescribed laws, at least within certain limits of pressure, than they should; when the pressure is increased they move closer together, but not quite as much, they allow the diminution of space to be somewhat smaller than corresponds to the increase in pressure. And they do this when, in the heating of a gaseous or vaporous body under constant pressure, the obligation having been imposed on the molecules of the body to travel their paths at greater distances {72} from each other than previously, and accordingly to occupy a greater space, one observes all too often that the molecules overenthusiastically array themselves at greater distances from each other and consequently allow the increase in space to be larger than is legally prescribed. The molecules appear to become aware of the fact that they have gone too far; or at least, the higher the temperature becomes, the more closely the behavior of the molecules of gaseous bodies approaches the correct one.

What is required of individuals in a molecular state that is said to exist in ideal perfection is not much; considered from our human standpoint, many things could be judged harshly. It is said that lascivious behavior between single molecules has never been known to happen; not even the inner feeling by which a molecule might feel drawn to another is permitted to be expressed or to attempt an action.

---

[187] I have not been able to trace this reference.

[188] This is a description of the deviation from ideal gas law properties, as exhibited in the behavior of real gases.

Just as little is permitted the inner feeling that could cause a molecule to behave repellently toward another. To be sure, as we have already frequently mentioned, repulsion between two molecules is very often given and taken; the repulsion that occurs as a consequence is said to be restricted to the purely superficial, and accordingly it happens that molecules are of a very elastic character.[189] That a molecule might behave boorishly improper by taking more space than it deserves is quite out of the question. It is said that the space that one molecule asks for itself as an individual, and so on for the entire space that the molecules forming a molecular state fill for themselves, is not allowed to be considered vis à vis the expansion of the state, indeed not even vis à vis the size of the average distance between the molecules. {73}

If we were citizens of such a state, a state that presumes the disregard and neglect of our important personal interests vis à vis the interest of the whole, and prescribes the strict observance of such constitutional provisions as the first obligation of a citizen, would we consider our existence as worthy of being human? No; and both of us know many who would do that even less than we. And the molecules? Selfless as they are, in their existence under such conditions they consider themselves as free. Human beings would not be capable of that, least of all those, who about 1840—at the time of the battle between the proponents of a certain philosophical school and their opponents, who fought it, not without reason, as leading to the negation of everything positive—sang to the lovely melody of Bertrand's "Departure" or a similar one from the old general:

Freedom is that which is adopted
As an individual into the general,[190]

although this would also be entirely appropriate for molecules in the so-called free state.

Many molecular states, those consisting of one kind of molecule, exist within very wide ranges of pressure and temperature at least approaching the ideal state, and they approach this state even more closely, as it happens, the lower the pressure becomes, and the higher the temperature.[191] But everything has its limits. So if for a gaseous body at a certain temperature the pressure crosses a certain limit, or if at a certain pressure the temperature falls below a certain limit, then it

---

[189] Tenets of the classical kinetic gas theory include the assumptions that molecules are perfectly elastic, and that that any forces between molecules act only at extremely close range.

[190] "Die Freiheit ist das in das Allgemeine / Als individuum aufgenommen sein." From "Spiessbürgers Freiheitslieder" (1843) by left revolutionary poet Georg Herwegh (1817–1875). "Bertrands Abschied" was the Germanized version of a patriotic tune by Friedrich Glück (1793–1840), portraying the leave-taking of Napoleon and his general Henri Bertrand on their way to St. Helena. The "certain philosophical school" referred to by Kopp was the leftist movement among Prussian intellectuals known as the "young Hegelians," which reached a high point about 1840.

[191] Divergence of real gases from ideal gas behavior is greatest at low temperature and / or high pressure, least at high temperature and / or low pressure.

becomes indolent[192] in the molecular state: if the pressure is raised still further, or the temperature further lowered, the diminution of volume that occurs no longer {74} approaches that which should occur as prescribed in the constitution, the average distance between the molecules is definitely closer than it should be, and perhaps also we might see the ganging up of molecules together; clearly, the public order in the gas-molecule citizenry is greatly disturbed. For different molecular states the limits where all this happens are quite different; for instance, for carbonic acid gas, law and order is still moderately maintained under certain conditions of temperature and pressure, while under the same conditions water vapor exhibits a definite absence thereof; and by the same token water vapor follows the fundamental laws in conditions under which formic or even acetic acid stubbornly refuse to do so, as we have earlier (p. {31}) described.

For the molecules of gaseous or vaporous bodies that are in this condition of refractoriness towards the law there are means to induce them to observe it: one gives them more to do, one raises the temperature, thereby compelling them to move faster, which they then carry out in an orderly way; or one reduces the pressure, forcing them further apart and allowing each to reflect and change its mind. With the tightening of discipline under whose influence the dissatisfaction of the gas molecules and their rebellion against the laws became noticeable, i.e., by the application of still higher pressure and still lower temperature, nothing is achieved for the reestablishment of public order in the molecular state; all this finally leads to another grouping of the elements of the state [Staat], and the establishment of a newly ordered state [Zustand] based on a different constitution than the one in force for the gas-molecule state. The new (so-called aggregate) form of the state [Staat] is usually at first the liquid form; but it can also be solid, or become so if the temperature lowering is carried far enough. {75} When the temperature of a gas-molecule state becomes so high that mere increase in pressure does not result in a transition to the liquid state, the relationships are however fruitless and uncertain; an extremely critical point for the existence of the organism of the state in its previous form, or perhaps in apparently its previous form, asserts itself; but we should not here and now go beyond this mere hint.[193]

[The World of Liquid Molecules]
It is well known that the transition of molecules from the gaseous state in which they previously found themselves to the liquid state can be effected in another manner than through the disciplinary measures just mentioned, namely by the so-called solution of molecules in liquids. This provides another unfortunate illustration how little chemists have known how to choose technical expressions that reflect the facts (the name for when a liquid is brought into another state by

---

[192] Kopp's word is "faul," which means not only lazy or indolent, but can also mean foul, bad or rotten—an intentional double meaning, for humorous effect.

[193] A critical state or critical point specifies the physical conditions of a multi-phase system under which the phase boundaries cease to exist. Kopp is of course making another double entendre.

union with a gas cannot be at all decently expressed[194]); such expressions are
suitably valued by humanists for that very reason. Not solution, no, subjugation is
what gas molecules are subjected to in this process; in order to keep our conscience as chemically pure as we can, let us name this using a somewhat more
appropriate word, absorption. It is always hard for gas molecules to be absorbed; in
fact it must be painful for them to go from the relatively free condition in which
they roamed here and there, to one in which they enjoy much less freedom of
movement, renouncing the sylphic or sylphid life[195] that they hitherto led.

Do we want to take a look at molecules after they have experienced such an
{76} alteration in their existence? The aerarium has an annex, to which this gate
allows us access. We can only take a very brief look, for we still have not yet come
to what is really interesting to see in the aerarium. And we ought scarcely expect to
see anything amusing in the annex. It serves also as an experiment station for the
scientific mistreatment of molecules, and terrible rumors circulate in the aerarium
concerning everything that is done to them.[196] But when it concerns cruel treatment of the inorganic world [von Unorganisirtem], we can look together upon that
treatment without real horror, even while experiencing compassion as such. So let
us just go in for a moment.

As we step through the darkened gate, we meditate to ourselves, looking at each
other, not *lasciate ogni speranza voi che 'ntrate*,[197] but rather think about the
nature of molecules of liquids.

The molecules of a liquid body, those identical smallest parts that compose a
perceptible quantity of a liquid body, are probably combinations of a larger
number of atoms than the gas molecules of the same body. There is much that
conduces us to assume this as probable, but to go into this in more detail would not
be prudent, considering both the time and the place, and for the same reason we do
not tarry to consider whether or not this or that fact contradicts or appears to
contradict this assumption, or how such a fact might perhaps nevertheless be
interpreted as standing in harmony with this assumption.

In any case, a molecule of liquid, however it is composed, moves as a unit,
while the characteristic intramolecular movements of the parts composing it are
likewise permitted. {77} The movements that the molecules of the liquid carry out
as a unit consist in vibrations. The molecules of a liquid body vibrate in various
directions around each other, but without overstepping a certain distance

---

[194] Kopp's little witticism refers to the circumstance that the German word "Verflüchtigung"
(volatilization) sounds much like "Verfluchung" (curse).

[195] "Sylphen- bez.-w. Sylphiden-Leben." The word "sylph" was a (probably arbitrary) coinage
by Paracelsus to denote an airy spirit.

[196] The agricultural experiment station (Versuchsstation) movement began, under the influence
of Justus Liebig, with the establishment of a station at Möckern, near Leipzig, in 1851. By the
time Kopp was writing there were over two dozen such stations in Germany, and the movement
had already spread to the United States. The quip about mistreatment is a lurid but obviously
innocently coincidental foreshadowing of the Nazi period.

[197] "Abandon all hope, ye who enter in." From Canto III, line 9 of Dante's *Commedia divina*.

separating each from the other; every molecule is bound to every other, around which it vibrates as it were by an invisible chain, which permits the first to move in all possible directions, but not exceeding that limit of approach. With rising temperature the separation distance between the molecules continually increases, the chain is lengthened, and the velocity with which the molecules vibrate increases; also the intramolecular motions of the parts composing each molecule are more lively and swift at a higher temperature than at a lower one.

If liquid molecules are composed of gas molecules,[198] the joining together of the latter to produce the former in many cases cannot possibly depend on bonding by handshakes, if the conception of the handedness of atoms presented here may still otherwise be applied. If—let us restrict ourselves to a few liquids, whose cruel treatment we will soon witness—a chlorine atom like a hydrogen atom is one-handed, a hydrogen chloride molecule, to whose existence the joining of the hands of the two atoms coming together is necessary, cannot seize another such gas molecule by offering this or that hand for the formation of each molecule of liquid; even less is this possible for one water vapor molecule to another, in which each has a single oxygen atom with its two hands tightly holding two hydrogen atoms; or for one ammonia gas molecule to another, in which each has a three-handed nitrogen atom associating with three hydrogen atoms by {78} the offering and mutual pressing of hands, and which as a consequence requires all its hands to remain united with the three latter atoms.

So in the liquefaction of hydrogen chloride, water, and ammonia by appropriate increase of pressure and decrease of temperature, the joining together of several gas molecules to a single new molecule does not depend on handshakes. But still, each such molecule of liquid must perform its vibrations with a velocity appropriate to the current temperature; that must be difficult for the constituent gas molecules within the larger structure, maneuvering without error while maintaining the legal separation between each other, since a handshake is impossible for them, and also probably it is impracticable to establish the requisite contact by each gas molecule slightly jostling its neighbor with its elbows. But: such a molecule of liquid *must* do this. Unfortunately, we know nothing about how it does it; the basis for the existence of so-called molecular compounds—those whose molecules consist of a determinate number of gas molecules that are not held together by so-called atomistic bonds, i.e., mutual holding of hands between two atoms—is unknown. The existence of such molecular compounds is nonetheless assumed, and indeed should be assumed in the current state of our knowledge (or lack thereof).[199]

---

[198] Here and in the following discussion, Kopp uses the phrase "gas molecules" to indicate not necessarily molecules in the physical state of a gas, but rather the *kind* of molecule that is normally present in the gaseous state, but that may also find itself in (e.g.) a liquid phase. For instance, in the case of hydrogen chloride (taken up in the next paragraph) the gas molecule is HCl, but the form of the hydrogen chloride molecule in aqueous solution, Kopp recognizes, may or may not be HCl itself; it could well be some combination of HCl with H_2O.

[199] See Kopp's discussion of molecular compounds above, pp. {46–51}.

And this not just in such cases as we just considered, where the so-called molecular union of several similar gas molecules is ordained as a molecule of liquid, but also in other cases, where dissimilar gas molecules must be united together to form a molecule of liquid in the same unknown fashion. So, for example, much favors the idea that in the absorption of hydrogen chloride gas in water, to the so-called aqueous (let's say, {79} more concentrated) hydrochloric acid, in addition to liquid molecules of water and chlorine, there must also be contained ones that are composed of hydrogen chloride gas and water vapor molecules combined together in a certain ratio. And similar things are to be assumed in not a small number of other instances.

In the current state of our understanding of liquid molecules we do not know how many gas molecules are contained in the liquid molecule, e.g. how many are in one liquid molecule of hydrogen chloride or water or ammonia; we have no idea of this. But the question might well be posed whether the liquid molecule of a body, when it is liquefied out of the gaseous state by appropriate increase of pressure and lowering of temperature, consists of just as many gas molecules, and whether it is in general of the same composition, as the liquid molecule of the same body when it is converted to the liquid state by absorption of its gas into a liquid.[200] We also do not know what numbers of gas molecules of one or another kind are joined together to a molecule of liquid when the latter consists of dissimilar gas molecules; in many cases we know at least the ratio between these numbers, but sometimes (for instance, for the just-mentioned liquid molecules consisting of hydrogen chloride gas molecules and water vapor molecules in aqueous hydrochloric acid) we do not even know that.

It would be of great importance to know all of this. But not knowing it is not as disadvantageous for considering chemical processes that concern liquid molecules as one might think. For such processes it is the gas molecules contained in the liquid molecules that always must bear the brunt, and therefore one can refer these processes almost entirely {80} to the gas molecules, without considering how they are grouped to form liquid molecules. And that is what one does; and that is what we intend to do now (in order to observe at least something in the annex, through which during the past remarks we have wandered, not in a straight path but through various corridors without particularly noting where we were going), by looking into that cabinet to see what's happening there.

[In the Annex]
Electrotherapeutic treatment of liquid molecules.[201] Good, we are in luck, that is just what we had wished to see. We are extremely interested in electrotherapy; we

---

[200] For example, is the hydrogen chloride molecule in liquid HCl—that which is created by liquefaction of gaseous HCl—the same as the hydrogen chloride molecule in aqueous solution?

[201] Kopp is reading a sign in the annex, informing visitors of the subject of the exhibit. Electrotherapy, or medical galvanism, was avidly pursued in the nineteenth century [27].

are convinced that it is advantageous for patients even in exquisite[202] cases. Only to electro-homeopathy have we not been able to warm, that subject for which the Russian whom we met two years ago in Ischia[203] was so enthusiastic, and who always came to breakfast carrying the French brochure on the subject.[204] We cannot look at everything that is happening here even in this one area, but at least we can see something!

This time, this something is not to be of the sort how the condition of liquid molecules can be ameliorated, but rather more like the assistance that the application of electricity provides for the diagnosis of the condition of such molecules: for the establishment of their true nature, and distinguishing what they are from what they appear to be. For the human world as well, electricity can be used with advantage in this regard; for the unmasking of the most crafty and obdurate malingerer, a gentle application of the electric brush leads her toward a confession. What we intend to consider here is of a simpler sort: how, {81} by applying a continuous electrical current through molecules of conducting liquids, the compound character of these molecules and of which elements they consist are revealed, and for this simple matter we want to remind ourselves in somewhat greater detail how that happens.

Electrolysis? you ask with a wry smile, and for a moment there returns to your face that pained expression of yours which has always indicated that you are worried about something. Electrolysis; but not the more difficult kind with glowing-hot molten compounds, but the very ordinary (because easily accomplished) kind, that from aqueous solutions.

In such an electrolysis an electric current brings about a decomposition of a hitherto stable compound. This decomposition is not always the sort which leads directly to the products, nor the sort which proceeds absolutely instantly for the compounds whose elements appear (as one said in that time when more worth than now was placed on precise diction).[205] What comes from these decompositions is often the result of intrigues that take place in the very heart of the liquid; not infrequently tricksters and schemers pursue their arts there, such as is portrayed in the rare (because they were intended only for the family) memoirs ("Some Events and Experiences" were recounted) by the venerable Kurhessian General

---

[202] "[A]uch ... in exquisiten Fällen": in German as well as English, "exquisite" is an obsolete and obscure medical term, meaning true or genuine, as opposed to spurious malingering.

[203] An island in the Gulf of Naples, about thirty kilometers from the mainland.

[204] Comte César Mattei, *Electro-homéopathie, principes d'une science nouvelle* (1879). Cesare Mattei was Italian, from Bologna; his electro-homeopathic institute was established in Geneva.

[205] The deliberate irony of the last phrase only makes sense when attached to the original German sentence, which is perfectly grammatical but highly convoluted: "Diese Zersetzung ist nicht immer eine unmittelbar die zu ihr gelangenden Verbindungen treffende, nicht eine für die Verbindung, deren Elemente zum Vorschein kommen, mit schlechthinniger Sofortigkeit vor sich gehende, wie man wohl in einer Zeit sagte, in welcher man auf präcise Sprachweise mehr Werth legte als jetzt."

("Feldherr") and Prime Minister ("Geschäftsvorsteher") M. E. von Schlieffen,[206] who died in retirement in 1825, and whose sympathies were so German that acting from conviction he avoided the use of foreign words to a degree that I have not been able to achieve, but which did not prevent him from having a different view of human affairs from that to which soon thereafter Schubart gave expression, in a verse of his still-touching poem "The Tomb of the Princes,"[207] in as much as at the end of 1775 he concluded for Landgrave {82} Friedrich II the contract with the representative of Great Britain according to which Britain would be given 12,000 Hessian soldiers to fight the rebels in the North American colonies, which much later he still believed he could defend.[208]

But what does any of this have to do with what there is to see in the aerarium, you ask? Well, now, you might have a point; let us leave this subject.

This much is quite certain: electrolytic decompositions are often not direct or primary, but rather only indirect or secondary. So one wishes to electrolyze water, for example. We find out that the purest possible water conducts electricity almost not at all, and so cannot easily be decomposed in this way. One adds to it a small amount of an acid, such as sulphuric acid, so that it will better conduct electricity. But the sulphuric acid does more than we really wanted from it, namely that it should simply give water a better ability to conduct electricity, and otherwise behave quietly; for it allows itself to be decomposed by the passing electric current, with the separation on the one side of free hydrogen, on the other of an atomic group which decomposes into free oxygen and a residue, which in turn, uniting with the elements of the still-present water, regenerates the original sulphuric acid, which apparently is still there, as if it had not clouded the water by the evolution of gas, as if the electricity had decomposed the water quite directly to the free elements, whereas in fact the decomposition of water, as we now know for certain, is only indirect. This game played by the sulphuric acid is continually repeated.[209]

---

[206] *Einige Betreffnisse und Erlebungen Martin Ernsts von Schlieffen* (1830). As Kopp remarks, Schlieffen (1732–1825) was a Hessian military and political figure, and also a writer. The archaic German expressions in parentheses were given by Kopp in quotation marks; I translate "Staats-Minister" as Prime Minister.

[207] Christian Friedrich Daniel Schubart (1739–1791), "Die Fürstengruft" (1774, published 1780), an extended attack on tyrannical rule.

[208] This monumentally grotesque sentence is another self-parody, as is revealed in the short paragraph that immediately follows. Here Kopp refers to the mercenary Hessian troops who fought on the British side in the American Revolutionary War. Landgraf (Count) Friedrich II (1720–1785) was the head of state of Hesse-Kassel (also known as Kurhessen or Electoral Hesse)—Kopp's homeland.

[209] This was the prevailing understanding of the role of sulphuric acid in promoting hydrolysis of water, at least since Faraday's work on electrolysis of 1834. It was thought that when the current starts, the electricity first splits H_2SO_4 molecules into $H_2 + SO_4$, then $SO_4 \rightarrow SO_2 + O_2$, with the elementary gases bubbling out of solution on each side. The acid is then regenerated when the SO_2 radical unites with two ambient water molecules, liberating a second molecule of hydrogen gas in the process. The overall process is thus $H_2SO_4 + 2H_2O \rightarrow 2H_2 + O_2 + H_2SO_4$. In a

How roguishly—speaking frankly and bluntly—does the sulphuric acid act! It is a slave to its predilection for electrical enthusiasm, thereby coming {83} entirely apart, or at least out of its true constitution, but always knowing how to get itself back together, rising like a phoenix (except in the wet way)[210]; the water must suffer the consequences, it must be decomposed all the way to its elements— further one cannot go—although it is entirely against its nature to suffer decomposition by the electric current. And we see this sort of thing quite often in electrolytic processes: what appears is the result not of primary but rather of secondary decomposition of the relevant compounds, and not infrequently it is uncertain whether the one or the other is occurring. But gradually we have had to accustom ourselves to judge by results, and especially in the present case we intend to permit ourselves to do that; in considering the electrolysis of various aqueous liquids, to pay attention only *to what* something, and not *how* this something, is decomposed.

As we approach the experiment bench we do not neglect to remind ourselves of something that is all too easy, due to its simplicity, to forget: that what one calls the electrical current are two currents of opposed electricities that flow through the conducting body in opposite directions, and that one calls the positive pole the end of the wire coming from the galvanic battery immersed in the liquid that supplies the so-called positive electricity, and one calls the negative pole the end of the immersed second wire coming from the battery that supplies the so-called negative electricity.[211]

The liquids on which some demonstrations will take place are concentrated hydrochloric acid, water (to which according to customary expression some sulphuric acid is added to improve electrical conductivity), and a concentrated absorption of ammonia gas in {84} water (in which for the same purpose some acid ammonium sulfate is dissolved). We observe each of these liquids with our mind's eye: the liquid molecules within appear to be somersaulting about, they are vibrating quite gaily. We will soon see the dancing, if not of the liquid molecules themselves, and also if not of the gas molecules composing the liquid molecules, but rather the dancing of the atoms composing the gas molecules, under the influence of the electricity.

(Footnote 209 continued)

Ph.D. dissertation submitted in 1884, Svante Arrhenius proposed, on the contrary, that the electrolytic current does not in fact initiate the dissociation; rather, electrically charged ions are permanently present in solution, and participate in the reactions as such. Arrhenius's theory of electrolytic dissociation appeared in the *Zeitschrift für physikalische Chemie* in 1887, five years after Kopp's book was published, and the theory was widely adopted within a few years.

[210] The phoenix was a mythical bird that arose reborn from its funeral pyre. "In the wet way" refers to the tables of chemical affinity of Etienne François Geoffroy and Torbern Bergmann in the eighteenth century, which distinguished between chemical affinity as exercised in dry reactions, as opposed to in solutions.

[211] Again, Kopp explains what was the prevailing understanding of electricity at this time.

Ad vocem[212] the dancing: we have already considered hydrogen as *generi masculini*,[213] and oxygen and nitrogen likewise are counted in the molecular world among the lords of creation, or the knaves of their ladies, those that have them; only chlorine belongs nominally to the so-called tender sex. So it is obvious that we have an incongruity: it won't work for dancing. Some light hydrogen atoms offer to dance as women—as happens also sometimes in the human world—for instance by wearing a distinguishing badge. The oxygen and nitrogen atoms gladly accept the proposal, but only the chlorine atoms, when the hydrogen atoms approach them to dance as if they were women, chlorine shows on this occasion once more that in her marriage she holds the inexpressible against hydrogen. Therefore, as we are to see it, all the hydrogens dance as women and all the other atoms as men.

The performance can finally begin. The electric current flows through the concentrated hydrochloric acid; hydrogen gas and chlorine gas, hitherto united with each other in hydrogen chloride, are separated from each other and appear in free form. But how remarkable: rather than being mixed with each other, on the positive pole only chlorine comes out from the liquid, and on the rather widely separated negative pole there appears only hydrogen gas. How {85} can that happen, that upon the separation of the chlorine and hydrogen atoms previously bound into a molecule of hydrogen chloride, hence located in the same place, these atoms, or atoms that seem to be very much like them, can appear at places so distant from each other?

To gain some understanding of this matter, let us place ourselves so that we have the liquid between the two poles exactly in front of us, the negative pole on the left, the positive on the right. Now all we have to do is to observe carefully. We see between the two poles—or so it seems to me—a difficult-to-estimate number of hydrogen chloride gas molecules (we have agreed that we will only have reference to the gas molecules themselves, and not how these may be united together with similar or dissimilar molecules to form liquid molecules) have entered into a Scottish country dance formation[214]; all chlorine atoms stand in a row, turning their backs to us, the hydrogens across from them, grinning in a friendly way at us (they seem to enjoy dancing as women), also in a line; the hand of each chlorine atom resting in the hand of the hydrogen across from her. Before the influence of the electric current one hydrogen chloride molecule bows after the other; how many do this in a given time depends on the strength of the current.

We keep one eye on the hydrogen chloride molecule that is closest to the left-hand or negative pole. There it is bowing; in the same instant the hitherto joined hands of the chlorine and hydrogen are released, the first atom says to the second,

---

[212] With regard to.

[213] Of the masculine gender.

[214] This is what is called a longways set in English and Scottish country dancing: all the men shoulder-to-shoulder in one line, all the women in an opposing line, with partners facing each other.

"Go to the left, let me go to the right," and in these two directions they in fact proceed apart. The hydrogen atom turns to the left toward the negative pole, and if it is not accepted by this one according to its nature, it has a mind to {86} leave the liquid; the chlorine atom turns to its right, and what it then does we will shortly see.[215]

With the other eye we look at the hydrogen chloride molecule closest to the right-hand or positive pole. It also bows; it behaves exactly as the worthy molecule we have just observed on the far left. The chlorine atom turns to the right, toward the positive pole, and if it cannot be used by the material of that pole to form a compound it remains single until further notice; the hydrogen atom turns to its left, and what becomes of him and of the previously mentioned right-turning separated chlorine atom, we shall now see.

These atoms that have turned toward the middle between the two poles do not proceed along a straight path to the other pole, originally so distant from them, in order to separate out of the liquid there; between the two poles there is nothing to see of a separated chlorine or hydrogen atom, and it is not at all clear why these two atoms wandering alone in opposite directions and meeting in the middle, should not unite together to a hydrogen chloride molecule.

How does it happen that a chlorine atom separating from a hydrogen chloride molecule at the left or negative pole can arrive at the right or positive pole, coming out of the liquid at its destination, without being identifiable along the way outside of a compound, and by the same token how does it happen that a hydrogen atom separated at the right or positive pole can sneak through to the left or negative pole? Or perhaps such an atom separated at one pole does not travel at all to the other pole, but rather a similar one comes in its place? These chlorine atoms, like these hydrogen atoms, are {87} indeed so similar as to be easily mistaken for each other. After these questions had already been treated by others, in 1838 a German chemist whom both of us highly esteem attempted to answer them in a very appealing way[216]; let us think of the matter, if not exactly in the same way as this scientist, at least similarly.

Turning to its right to the closest hydrogen chloride molecule, the chlorine atom separating from the hydrogen chloride molecule at the negative pole tears the hand of the hydrogen atom away from that of the chlorine atom, the latter then acts similarly to the hydrogen chloride molecule next door, and this process goes on

---

[215] The essential character of the process that Kopp describes here on pp. {85–88} was first proposed in 1806 by Theodor von Grotthuss (1785–1822), to provide an explanation of the mysterious circumstance that products of electrolysis appear only at the electrodes, and not throughout the solution [28]. Grotthuss suggested the kind of molecular dance depicted so anthropomorphically by Kopp; chains of decomposition occur throughout the solution, but radicals remain chemically bound to neighboring radicals, other than during the fleeting instants when molecular partners are exchanged. Consequently, liberated elements appear only at the ends of the chains, that is, at the poles. This theory was accepted, with variations, by most chemists and physicists until the late 1880s. See above, n. 209.

[216] I have not been able to identify this reference.

until finally the chlorine atom from the last undecomposed hydrogen chloride molecule at the right or positive pole, now robbed of its associated hydrogen atom, is left over, and knows nothing better to do than to depart from the liquid. Exactly the same sort of thing proceeds from the right-hand or positive pole immersed in the liquid: the hydrogen atom separating from the hydrogen chloride molecule in contact with this pole, turning to its left to the nearest hydrogen chloride molecule, demands the hand of the chlorine atom of this molecule; thus freed of its handhold, the hydrogen atom of the compound acts the same way to the nearest hydrogen chloride molecule to its left, and in this way ultimately a hydrogen atom remains in excess at the left or negative pole, and launches himself into the air.

All of this happens at lightning speed, indeed requires no time at all; in the same instant following our ideas, one hydrogen atom comes left to the negative pole out of the hydrogen chloride molecule that happened to be standing there, and a second hydrogen atom separates in consequence of the decomposition of {88} hydrogen chloride at the right or positive pole and the propagation of the just-described process from right to left through the entire column, the two hydrogen atoms joining hands and emerging from the liquid as a hydrogen molecule. In exactly the same way two chlorine atoms separate out virtually simultaneously at the positive pole, associating together to form a chlorine molecule emerging from the liquid. This repeats itself in an unimaginably short time period; in the place of the undecomposed hydrogen chloride molecule (which one might say has gone to waste under the influence of its electrical enthusiasm) there immediately emerge, from the large number of hydrogen chloride molecules that previously stood outside the column, new ones within it; in the direction from left to right chlorine atoms join together, and from right to left hydrogen atoms join together continuously into molecular couples.

The above-mentioned scientist has depicted this whole process somewhat differently and accordingly has thought of it in a somewhat less simple and perhaps more correct fashion; we are not stubborn, and gladly concede that the motions of the individual atoms, due both to their tiny size of the one and the swiftness of the other, cannot be established by direct observation. According to this scientist the molecules approach each other between the two poles (let us continue to say that the negative pole is immersed in the fluid to the left, the positive to our right) not, as we assumed, in the form of a Scottish country dance formation, but rather in the form of a minuet; not such that all the atoms dancing the part of boys stand across from those dancing as girls, but such that everyone, the former as well as the latter, forms a single line; the electric current, he says, exerts a rectilinear force on the atoms between the poles, through which the latter type of their arrangement {89} is determined as necessary. We imagine all the hydrogen chloride molecules placed accordingly between the poles, and in fact positioned so that in each one the hydrogen atom is directed toward the left-hand or negative pole and the chlorine atom toward the right-hand or positive pole. Thus a hydrogen atom stands to the left at the negative pole, a chlorine atom stands to the right at the positive pole. Under the influence of the electric current that hydrogen atom separates out at the negative pole, this chorine atom separates at the positive pole. The chlorine atom left in the lurch at the left now seizes the hand of the hydrogen atom in the closest hydrogen chloride molecule to the right, whose chlorine atom hurries further on to the right, and this exchange of handholds is propagated to the right throughout the whole line, so that each chlorine atom lets go of the hand of the hydrogen atom standing next to her on the left and seizes the hand of the hydrogen atom standing next to her on the right. That proceeding from the right, where the hydrogen atom abandoned by the chlorine atom seeks lodging and consolation with the chlorine atom of the closest hydrogen chloride molecule on the left, to the creation of the new atomic arrangement, it scarcely needs to be said that both kinds of atoms in the hydrogen chloride molecular couple powerfully take part.[217]

Is the dance now over? No. For in consequence of the aforesaid rectilinear force, all of the newly arranged hydrogen chloride molecules, the hydrogen atom of each being directed to the right and the chlorine atom directed to the left, now turn in a half-circle as in military formation, in consequence of which the hydrogen atom in each hydrogen chloride molecule now faces left, and the chlorine atom faces right. Then everything repeats as has already been described, and this goes on further; the hydrogen atoms contained in the hydrogen chloride molecules originally to the right, or the hydrogen atoms emerging from hydrogen chloride molecules newly appearing from the bystanders in {90} place of those that were decomposed, proceed through the molecule in wavy paths to the left until they reach the negative pole, where they separate out, and the chlorine atoms proceed in a corresponding way to the right until they reach the positive pole. Or on the whole, as this has been expressed pithily and to the point: the hydrogen and chlorine atoms perform a grand chain together, as in a French courtly dance.

As we said: the matter may well be like this. But we will not tarry to consider what may speak in favor of this or the other conception. For we have already stopped all too long to consider such a factually simple process as the electrolysis of concentrated hydrochloric acid. And before leaving this room we still want also to look at what has now been prepared for us. The simultaneous decomposition of several compounds by one and the same electric current, and all of the remarkable features of this, is to be demonstrated to us. But we wish to employ the considerations connected with the perceptible factual matters from the viewpoint of the first of the two briefly discussed conceptions, namely from the point of view of the simpler one.

[The Dance of Electrolysis]

There we have in three containers—*cella*, cells are what the somewhat Abruzzan-looking attendant[218] calls them—the three previously named liquids, prepared

---

[217] This corresponds to the starting position of a "grande chaîne" in a minuet: couples form a single line with partners facing each other, then partners progress down the line in opposite directions using alternate hands, all men moving in one direction, women the other. See the end of the next paragraph.

[218] Kopp's phrase "etwas abruzzenmässig aussehenden" = "in appearance somewhat Abruzzan." Abruzzo is a mountainous and predominantly rural region of east central Italy, historically associated with Naples.

such that it is obvious that something is about to happen with them: to our left once again concentrated hydrochloric acid; in the middle, water acidified with a little sulphuric acid; and to the right, a concentrated aqueous solution of ammonia gas with a little ammonium sulfate added. The negative electricity comes from the {91} battery on our left, first into the hydrochloric acid, it flows through it, comes out through a bent wire and thence into the water; after flowing through the latter it goes again through a wire to the aqueous ammonia, and after passing through that it finally turns right and returns through a wire to the battery; by the same token, the positive electricity coming from the battery goes first to the right to the aqueous ammonia, from this to the water, then to the hydrochloric acid, and finally exits from the latter to return to the battery. Thus, in each cell the positive pole is immersed on the right side, the negative pole on the left. When the current flows, decomposition occurs in each liquid; directly or indirectly, the hydrogen chloride, water, and ammonia molecules are reduced to their elements. The liquids are already saturated under current conditions with the gases that are freed during the decomposition. The strength of the current is always the same along its entire path; the greater resistance exerted by the progress of the current at a given point, i.e., in one of these liquids, distributes its effects equally over the whole pathway.

Just now the battery is to be activated, allowing the electric current free rein. Now it is even more important than earlier to observe not just with a sharper eye, but, so to speak, with a mind's eye multiplied many times over; now it is also a matter of counting in all three cells the molecules, atoms, and atomic hands in the shortest conceivable instant of time, or one even shorter than that. It is good that there are two of us; we can check our counts against each other to see if they agree. Using a specially constructed apparatus, the attendant allows the electrical current {92} to operate on the liquids, but just for a moment.

What did we just see? Between the poles immersed in each of the liquids, the molecules that were about to be decomposed immediately assumed the formation of a Scottish country dance: the hydrogen chloride molecules in the previously described way, so that all the chlorine atoms stood in a line, each holding in its one hand the hand of the hydrogen atom standing across from it; the water molecules such that all the oxygen atoms stood in a row and each had seized with its two hands the two hands of the two hydrogen atoms standing across from it; the ammonia molecules such that all the nitrogen atoms stood in a row and each had grasped with its three hands the three hands of the three hydrogen atoms who stood across from it. Virtually simultaneously the decomposition took place; with the most intense concentration but also with results that agreed with each other, we counted how many molecules within the same short time interval experienced decomposition, how many firm hand-holds were loosed, how many atoms of one or another kind were separated out. In the hydrochloric acid 600 hydrogen chloride molecules were decomposed, 600 hand-holds loosed, 600 hydrogen atoms separated out to the left at the negative pole, and 600 chlorine atoms separated out to the right at the positive pole. In the water 300 water molecules were decomposed, 600 hand-holds loosed, 600 hydrogen atoms separated out to the left at the negative pole, and 300 oxygen atoms separated out to the right at the positive pole.

Finally, in the ammonia solution 200 ammonia molecules were decomposed, 600 hand-holds loosed, 600 hydrogen atoms separated out to the left at the negative pole, and 200 nitrogen atoms separated out to the right at the positive pole.[219]

What a lovely agreement—despite the very different numbers of the various substances that were decomposed—namely, {93} that from each of these compounds the same number of hydrogen atoms were separated out at the negative pole; but even more important for us is to take notice on what this agreement rests: that in general the same amount of current electricity working to decompose various compounds looses an equally large number of hand-holds. And this happens, too, whether one considers the decompositions to be direct or indirect. For in the latter case, in order that a certain number of hand-holds between the elementary atoms that actually leave the liquid are loosed, it is first necessary to loose an equally large number of hand-holds between other atoms or atomic groups, and these then operate upon the molecules of the compound that is ultimately perceived to undergo decomposition.

As we have already learned today in an earlier hour, the elementary atoms separating from the molecules to the left and to the right in each of the decomposition cells do not do so individually. In each of the cells the same pattern is followed—by the hydrogen atoms dancing as girls and separating out at the negative pole, and by the chlorine, oxygen, and nitrogen atoms dancing as boys and separating out at the positive pole—which can be found in the illustrated catalog of J. C. Schmidt's admirable factory for cotillion supplies in Erfurt,[220] describing the dance instruction at the conclusion of a new cotillion tour, when all the women are brought to the one side, and all the men to the other: "Women join hands to form couples and chassé[221] down to the left; men likewise join hands to form couples and chassé down to the right." In our case, from each cell 300 hydrogen atom molecular couples chasséd {94} to the left into freedom, joined one-handedly; to the right 300 chlorine atom molecular couples joined one-handedly from the hydrochloric acid cell, 150 oxygen atom molecular couples joined two-handedly from the water cell, and 100 nitrogen atom molecular couples joined three-handedly from the ammonia cell. Whereby I will not neglect to note that if, as a matter of duty, the factory literature for cotillion supplies is heeded by chemists, so also the said manufacturers should, for their own good as well as for the wider benefit of the dance-loving human world, pay a little attention to the chemical world; they could find there a surprising selection of beautiful patterns for cotillions. I might especially recommend, for those who are interested, the

---

[219] This is a formulation of Faraday's laws of electrolysis (1834), which posit that the weight of a substance deposited or freed at an electrode is proportional to the amount of current that has passed between the electrodes, and is also proportional to the respective equivalent weights of the reacting substances.

[220] A later edition of what is obviously the same catalog to which Kopp here refers is *Preisbuch über Cotillon- Ball- und Scherzartikel, Saaldekorationen, Sommerfestartikel, u.s.w.* (Erfurt: J.C. Schmidt, 1911).

[221] A fast sideways sliding step done as a couple.

instructive booklet on this subject: "Chemical Formulae of Minerals in Geometric Figures," written by P. Sigmund Fellöcker (Linz, 1879).[222]

But this was really hard work, with all that counting, and nothing like a recreation. Hopefully we will find compensation in the rooms that we have not yet visited, which have been described to us as offering so much pleasure. To visit these rooms we had best leave the annex and return to the main building of the aerarium.

[The World of Hydrated Salt Solutions]

We inquire of the somewhat grim-looking attendant—who would gladly have shown us more—regarding the shortest way back to the main building. The attendant accompanies us through a gloomy corridor. To show courtesy in taking our leave we exchange a few words with him; we ask him what is behind a small door by which we are walking. We receive as an answer: they are salt solutions that have been discarded (*rigettate*) and cleared out of the walkways. Oh, please let us look in for a moment, I ask; I am always interested in these solutions and their compositions, and I vaguely remember {95} having published something on this subject around the year 1840.[223] Polite as you are, you yield to the request, and we enter into the very hot room.

There they are, in very dusty bottles—in Naples, too, not everything is kept perfectly clean; even the attendant comments, *un poco sporco?*[224]—the favorites of my youth and frequent companions of my maturer days; there they are, the saturated aqueous solutions of zinc vitriol,[225] of the bitter salt that often also is used outside of our professional circle,[226] of the similarly-oriented salutary preparation that Glauber rightly called the *sal mirabile*,[227] and many other salts. How well do they feel? The Glauber's salt in solution seems not to be doing very well in the high temperature here, over 34°; it has thrown off some anhydrous sodium sulfate, which has accumulated on the bottom of the bottle. What might be wrong with the dear hydrated salt? I believe it is suffering an attack of liquid dissociation.[228]

---

[222] P. Sigmund Fellöcker, *Die chemischen Formeln der Mineralien in geometrischen Figuren* (1879).

[223] By studying solubility regularities of saturated aqueous salt solutions using mixed salts in selected pairs, Kopp argued in an 1840 paper [29, 270] that certain pairs of compounds must be forming double salts in solution, which decompose back into single salts when the solvent is removed. This is an example, Kopp noted, how solubility studies can reveal the existence of compounds "that are not easy to detect in other ways."

[224] A little dirty?

[225] Zinc vitriol or white vitriol is zinc sulfate, used in tanning, dyeing, and pharmacy.

[226] German Bittersalz, English Epsom salt, i.e., magnesium sulfate.

[227] Glauber's "miraculous salt" is sodium sulfate.

[228] Sodium sulfate has unusual solubility characteristics. With rising temperature, the solubility of the salt in water increases, but only until 32.4° C. is reached; at this point the salt begins to become *less* soluble with increasing temperature. Hence a precipitate begins to form when the temperature of a saturated solution of sodium sulfate is raised to about 34°. Kopp was aware that

Let us, contemplating this process for only for a very short time, remind ourselves of past and present views on the so-called constitution of aqueous solutions
of hydrated salts: how strongly there comes upon us the comparison to *Sonst und
Jetzt!*[229] Forty years ago we thought we fully understood what kinds of bodies or
molecules are present in such a solution mixed together. Until shortly before that
time reigned the conception that in every aqueous salt solution, water is the
dissolving liquid, as opposed to the anhydrous salt which is the solute, whether the
solution is that of a known anhydrous salt or that of a hydrated salt. Salt hydrates:
compounds of anhydrous {96} salt with water, according to fixed proportions; salts
containing so-called water of crystallization were long known. That an anhydrous
salt unites with water to form such a compound was considered, according to the
electrochemical theory that became accepted in the second decade of our century,
to be based upon a balancing of the opposite electricities attached to the components of the compound; the electricity was supposed to have displayed the marriage certificate, as for all compounds that exhibited fixed proportions, so likewise
for this kind, which by the way was never actually shown. But then—it was
probably Gay-Lussac who first presented another idea to chemists in 1839[230]—it
was thought that a certain amount of water itself can be contained as part of the
solute in the solution, that a hydrated salt can be dissolved as such in a further
quantity of water, in which case an anhydrous salt in solution would need to be
recognized as united with a certain portion of the water according to a fixed
proportion to the dissolved hydrate, and this hydrate would need to be recognized
as united with the remaining water according to a variable proportion, to form a
homogeneous liquid.

Judging from the expressions commonly used, this conception is probably the
most generally accepted one today; but it is to be sure a somewhat outdated
framework—constituting a part of the theoretical structure of chemistry as it now
exists—that presents itself as the continued presence of molecular compounds in
solution.[231] According to this conception, there remains together in solution,
at least at first, what was united, or comes together, in the crystalline state, which is
determined for each other: salt and water according to fixed proportions,
depending namely if a salt with water of hydration is dissolved in water, or an
anhydrous salt that is capable of forming a hydrate under the circumstances in
which the {97} solution occurs. If the hydrated salt in solution has formed from the

---

(Footnote 228 continued)
this is because the decahydrate decomposes at this temperature, leaving the less soluble anhydrous salt.

[229] A reference to Kopp's popular lecture, *Sonst und Jetzt in der Chemie, ein populär-
wissenschaftlicher Vortrag* (Braunschweig: Vieweg, 1867), which compared "Past and Present"
understandings of chemical substances.

[230] Gay-Lussac [30, 1012–13].

[231] Here and in the following passages Kopp suggests that the attraction between the molecule of
anhydrous salt and the molecules of water of crystallization can be considered a kind of
"molecular compound" similar to those phenomena discussed on pp. {46–51} and {78}.

anhydrous salt and the appropriate allowance of water according to fixed proportions—if it was not supplied at the beginning—and has united with the remaining water of solution in an accommodating fashion to form a homogeneous liquid, if all which belongs together, is together, then we have pre-1848[232] quiet, like that which Dingelstedt's cosmopolitan night-watchman in that day (1842) expressed in song, or recommended:

> So do not toss and turn in ill temper
> In your bed, and do not lie awake there;
> But quietly draw your nightcap,
> Satisfied, over your countenance.
> The dog in his stall, the man at home with his wife,
> The maiden with her beau, as right and duty allows,
> So rest, and by no means stir,
> For no harm shall come to the town.[233]

According to this conception, tranquility rules also in a liquid in which decomposition of a dissolved hydrate of a salt begins when warmed to a certain temperature, and the decomposition goes further when the temperature is raised even higher, as soon as the latter has become stationary. In an aqueous solution of sodium sulfate up to 34° one must view the hydrate of this salt, called Glauber's salt, as the solute; if a 34° saturated solution is heated further, a poorly-soluble anhydrous salt precipitates, and in the liquid one must assume that a portion of the quantity of hydrate that was present at 34° is still there, the other portion having been precipitated to the anhydrous salt, its water content transferring to the water itself. But whenever the temperature is constant, both equilibrium and rest are established: the water molecules apportioned to the anhydrous salt to form the still-dissolved hydrate remain quiet regarding the latter, and do not attempt {98} to emancipate themselves, as it were, into the free water of the solution; the molecules of the water of the solution remain as they were, and do not attempt to change themselves into water of hydration; everything remains as pre-ordained, such that the equilibrium is established corresponding to the temperature.

But what a change in the accepted view have we had to experience! As I understand the matter, we don't know anything more than we ever did why molecules of anhydrous salts unite with water molecules according to fixed proportions to form molecules of solid hydrated salts. At one time the reason was supposed to be the equalization of opposite electricities, and whether that is true or

---

[232] The German word is "vormärzlich," meaning pertaining to the era that preceded the insurrections of March 1848, presaging the (ultimately failed) liberal revolution of that year. German historians still use the terms "Vormärz" and "Nachmärz" to designate the eras adjoining that eventful month.

[233] "Auch wackelt nicht im bösen Willen / An Eu'rem Bett und räkelt nicht; / Die Zipfelmütze zieht im Stillen / Zufrieden über's Angesicht. / Der Hund im Stall, der Mann beim Weibe, / Die Magd beim Knecht, wie Recht und Pflicht, / So ruht und rührt Euch nicht beileibe, / Auf dass der Stadt kein Schad' geschicht." Franz von Dingelstedt (1814–1881), *Lieder eines kosmopolitischen Nachtwächter* (Hamburg: Hoffmann, 1842), "Nachtwächters Stilleben."

not was very much an open question. Today, when these compounds are placed in the same drawer as those denominated "molecular compounds," a drawer which holds everything connected with what is considered to be the cause of the cohesion of any molecule that cannot be fit into any other pigeonhole of the relevant section of the chemical filing system, today one cites no specific reason at all—apparently not because of mere shyness, or because it is too self-evident—and this circumstance—namely, that no one gives any reason other than one that simply restates the facts—can surely not remain an open question, but is true.[234]

We contemplate what we understand to be a molecule of a solid body: that—corresponding precisely to what we remember from earlier (pp. {76} ff.) regarding the molecule of liquid bodies—it is one of the identical smallest parts of which a perceptible quantity of a solid body consists, a unification of a number of atoms that vibrate as a whole. The vibrations of the molecules of solid bodies occur {99} in a different way from those of liquid or gaseous bodies. The molecules of solid bodies oscillate around fixed equilibrium positions; as if set on a short chain whose last link can glide back and forth on a rod of a certain length, each molecule moves in front of another, while also spinning and turning. The molecules of solid bodies can also be viewed as consisting of gas molecules, or, if we are not dealing with volatile bodies, as consisting of atomic groups that correspond to gas molecules to the extent that they too are the smallest atomic groups that suffer changes through chemical reactions. We know nothing about the number of gas molecules—we will employ only this simpler expression—that are combined to form one molecule of this or that body, we know nothing, for instance, of how many water vapor molecules are in a single molecule of solid water. We also do not know the basis for the cohesion of gas molecules to form a molecule of a solid body, just as little do we know the cause of cohesion of several gas molecules to a single molecule of a liquid, as we discussed earlier (p. {77} f.); which is commonly expressed by saying that also in the first case there occur molecular bonds between gas molecules.

The molecules of a solid body do not need to be composed only of similar gas molecules, they can also be composed of dissimilar ones combined in fixed proportions; so, for instance, the solid compound of chlorine with water formed by the introduction of chlorine gas into water at 0°; again, we do not know the numbers of gas molecules that form a single molecule of this compound, only the ratio, that to each molecule of chlorine gas come ten molecules of water vapor; and just as little as the cohesion of similar molecules can we imagine that of dissimilar ones as depending on hand-holds, {100} or on whatever other

---

[234] Kopp is pondering the fact that certain anhydrous salts, the atoms of which have already satisfied all their valence bonds, unite with definite numbers of water molecules in an apparently chemical fashion, to form hydrated salts (salts containing water of crystallization). These hydrated salts then dissolve as such, in larger amounts of solvent water molecules. Then, under certain conditions, the salt molecules apparently lose their water of crystallization, and dissolve (or precipitate) as the anhydrous salt molecules. He notes here that such facts were not theoretically explicable in his day.

precise notion one might cite. And the matter is just the same—not to mention many other compounds—with hydrated salts, which consist of an anhydrous salt (which, as we mentioned earlier, should be viewed as corresponding to a gas molecule) and water, according to one or another fixed proportion, and which are also molecular compounds.

They are quite a crew, these molecular compounds as considered in the light of recent ideas. But in what relationship are the components of these compounds found? Earlier, in considering the molecules of liquids, we were concerned to call the thing by its correct name, but one prefers to close one's eye to the issue as long as one can. But now, when the matter comes to our attention once more and in this fashion, this won't do any more. The components of a molecular compound have associated with each other not by a handshake, an act that in the molecular world would make a compound respectable. But in what relationship do the molecules coming together into a molecular compound stand with one another? I regret to have to say it: they live in convibration.[235]

It is to be expected that such compounds do not prove to be more permanent, and experience confirms this expectation all too clearly. Molecular compounds whose components are volatilizable—such as the just-mentioned compound of chlorine with water—are decomposed in the attempt to volatilize them—to a mixture of chlorine gas and water vapor. It would be an exception to the rule if, in the case considered this morning (p. {46} f.), the formation of a molecular compound that could exist undecomposed in the volatile state, even if only partially, were to be recognized; but on that occasion we already intimated that the actual occurrence perhaps could be explained in a different manner: let us say {101} now more clearly, that one might have some reasons to doubt whether a compound of dimethyl ether actually forms with the elements of hydrogen chloride. If such molecular compounds as hydrated salts are dissolved in water, it is consistent with what we now know to assume that frequently—not always—they are still preserved entire in the solution up to a certain maximum temperature: that the salt and water molecules that have joined together to form the solid compound still vibrate as a whole, where these molecules and the atoms inside them still have their own characteristic motions; above this maximum temperature the characteristic motions of the simpler molecules, the molecules of salt and water, become so lively that the more complex molecule, that of the molecular compound, cannot maintain itself in its previous existence, so that some or all of the water molecules depart from it, associate themselves with the molecules of the water in solution, and now, as the latter do, carry out their vibrations in the free state. This decomposition of the original hydrated salt in water to form an anhydrous salt (or a hydrate with less water) and free water happens at first only to relatively few of the molecules of the original hydrated salt, but with increasing temperature ever more

---

[235] "Aber in welchem Verhältniss zu einander stehen denn die ... Molecüle? ... [S]ie leben im Convibrat." Since "Verhältnis" means ratio or proportion as well as relationship, this is another double entendre. Convibrat is a nonce word, intended humorously to evoke the connotation of cohabitation.

of them; for each temperature, an equilibrium is established between the amounts of water of the solution and water of the solute. The process is so similar to that considered earlier today (p. {56} ff.) of the dissociation of a gaseous compound, that it has been called the dissociation of a liquefied compound or of a liquid, or sometimes as liquid dissociation for short.

The more recent conception that has just been sketched has very much going for it, even if—we cannot here avoid being mindful of this—in some {102} respects it would be very desirable if we could make for ourselves a somewhat more precise idea concerning it. The more recent conception corresponds as well as it can at the moment to what is recognized as the basis for the phenomena of heat; it is consistent with liquid and solid bodies, as they are conceived of, and with that which has been investigated with greater certainty due to its feasibility, namely with gaseous bodies, and it lends itself to beautiful comparisons of molecules with organisms, and to the proposing of instructive observations about the battle for existence among the molecules. It teaches us much of great importance, but unfortunately also how chaotic the conditions are in the solution of a hydrated salt with all of its possible dissociations. As far as the high temperature of this room permits, we still want to give the subject some more attention.

We can do that in just a few minutes. After all, the foundation for this is in our recent memory, from the consideration on what the dissociation of a gaseous compound depends, and what occurs when it happens; what happens in the dissociation of liquids is entirely analogous. For each temperature above that at which dissociation begins, an equilibrium proportion is established between the quantities of liquid of solution, of the yet-undecomposed hydrated salt, and of that other material besides water that results in consequence of the beginning of decomposition of the salt, if this other material remains in solution. But a state of rest as well? O! to believe that everything in the liquid is arranged lock-step and in repose—that is an outdated concept, which became untenable when, in the March-days of the science, springtime broke out for chemical theory too (according to the calendar followed by a portion of the human world, it was in the late summer of 1850), on the broader basis of the motion of the smallest parts of the components of compounds.[236] {103}

Let us fix our eye on a most simple case: let a hydrated salt, which for every one molecule of the anhydrous salt there is one molecule of water, undergo aqueous solution, in which a constant exchange of places, or dissociation, of the individual parts of the liquid takes place; let the temperature at which this happens be stationary, i.e., let us assume to be constant the average temperature of the various somewhat unequally warm spots of the space filled with the liquid. At a slightly warmer spot a slightly larger number of the hydrated salt molecules that find themselves there (or that can be imagined there) are decomposed to water molecules and molecules of the anhydrous salt—for short, we shall call them salt

---

[236] With the expression "March-days" ("Märztage") Kopp is once more playing on the idea of revolution (see above, n. 232), but this time a revolution in science that (he is suggesting) occurred in the late summer of 1850. He is almost certainly referring to Alexander Williamson's ground-breaking work on ethers. See Rocke [13, 1–3].

molecules; immediately after the arrival of this part of the liquid or a portion thereof at a slightly less warm spot a certain number of water molecules must again get involved with an equal number of salt molecules in molecular combination, to re-establish the equilibrium proportion. The latter molecules of both kinds unite, just as merest chance leads them together, one to the one and one to the other kind; perhaps they are happy about this, but scarcely do they think that their joy is about to end. The newly formed salt hydrate molecules come once more to a slightly warmer spot, and the molecules composing them separate from one another again just as thoughtlessly as they had united; the joy of being bound together was very short-lived. But this does not condemn each of the separated molecules, salt and water, to an unforgettable and sorrowful solitary life, but, as circumstances bring it to a slightly less warm spot, each enters once again into a volatile (not in the physical sense) combination with the first molecule of the other kind that crosses its path. Each of these molecules behaves with respect to a molecule of the other sort as fortune does with respect to humankind, concerning which Heine in the *Romanzero* taught the lesson: {**104**}

> Fortune is a most fickle lass,
> Who tarries not gladly many a day;
> She strokes your hair gently and smiles in the glass,
> Kisses you quickly, and flitters away.[237]

That must be quite a business, worse than in a Fourier phalanstery![238] But in all this chaos there is one thing in order, which nature, like an energetic commissioner of the ball, supports: that under the same external circumstances, i.e., at constant temperature and of course at constant concentration of the solution in which certain numbers of simpler molecules (e.g. of salt or of water) are present, a quite distinct number of molecularly bound molecular couples may, or must, be present, but—as is sufficiently obvious from the previous discussion—not definitely engaged, but rather in a constant exchange of partners, whereby each associates for a time with a molecule of another kind, floats around with him for a while, then having been let loose, buzzes around alone, then again enters into a short association with another one of the other kind; only the number of couples and the ratio of that number to the number of molecules vibrating alone must remain constant.

And where all this takes place, where the relationships between the molecules of the one and the other sort are so frivolously entered into and then once more abandoned—chemists speak of this, that in solutions in which hydrated salts undergo dissociation, there exist, at one or another temperature, certain relative numbers of molecules of hydrated salts joined together from salt molecules and water molecules according to fixed proportions. Oh, these chemists!

---

[237] "Das Glück ist eine leichte Dirne, / Und weilt nicht gern am selben Ort; / Sie streicht das Haar dir von der Stirne, / Und küsst dich rasch und flattert fort." Heinrich Heine (1797–1856), *Romanzero* (1851).

[238] A phalanstère was the projected (but never built) physical heart of the sort of utopian community prescribed and designed by the socialist Charles Fourier.

What would be the simplest thing, and what would correspond to the arrangements of the races of mankind that are more inclined toward higher civilization, would be to assume that each single molecule of one kind might unite molecularly with a single molecule of the other kind. But according to recent views, the processes of dissociation in solutions of hydrated salts are truly Turkish,[239] {105} which suggest that to one molecule of salt several molecules of water are present: e.g., ten as in the hydrate of sodium carbonate obtained by crystallization of common soda, or the hydrate of sodium sulfate produced from actual Glauber's salt in crystals. In the dissociation (mentioned above) of Glauber's salt in saturated solution that happens at 34°, all ten water molecules previously bound in a molecular arrangement with each molecule of sodium sulfate run away; is it any wonder that the sorely vexed (because ten-times betrayed) anhydrous salt molecule that is left behind thinks: I'm not going to play along any more, and resentfully precipitates out of solution?[240] It goes somewhat less badly for the soda, whose dissociation in saturated solution also begins at 34°,[241] but nonetheless not at all well; on this occasion each molecule of sodium carbonate is abandoned by nine water molecules with which it previously ran around in molecular combination; only one water molecule remains faithful to him, and at a higher temperature, molecularly bound to this one, preferring a solid state, it withdraws from the hurly-burly of liquid molecules and retreats back into insolubility.

[Conclusion]
But we too had better retreat back into the main building of the aerarium, for we cannot bear any longer the heat in this room.

Now on to see more interesting things! What? It's already one o'clock? Really. We wanted to be at Hassler's[242] already at this hour. So let's go. But if it can be arranged, we certainly want to come back here again.

References to *From the Molecular World*

1. Roscoe HE (1900) Bunsen Memorial Lecture. J. Chem. Soc. 77:513–54
2. Roscoe HE (1919) Ein Leben der Arbeit. Akademische Verlagsgesellschaft, Leipzig
3. Baedecker K (1883) Italy, Handbook for Travelers, Pt. 3: Southern Italy and Sicily, 8 ed. Baedeker, Leipzig

---

[239] A reference to polygamy. To a nineteenth-century German, the Turks would not be counted among the "höherer Civilisation zugeneigteren Menschen-Völker."

[240] See above, n. 228.

[241] The modern determination of this transition temperature is 35.4°. At this temperature Kopp knew that the decahydrate of sodium carbonate becomes a monohydrate, with slightly different solubility characteristics.

[242] Not identified; possibly a restaurant or hotel.

4. Clausius R (1857) Ueber die Art der Bewegung, welche wir Wärme nennen. Ann. Phys. 100:353–80
5. Fownes G (1878) A Manual of Elementary Chemistry, 12th ed. Lea, Philadelphia
6. Partington JR (1961–64) A History of Chemistry, 4 vols. Macmillan, London
7. Graham T (1849) On the Motion of Gases. Phil. Trans. Roy. Soc. 136:573–631, 139:349–91
8. Graham T (1833) On the Law of the Diffusion of Gases. Phil. Mag. 2:175–90, 269–76, 351–58
9. Partington JR (1949) An Advanced Treatise on Physical Chemistry. Wiley, New York
10. Mason EA, Kronstadt B (1967) Graham's Laws of Diffusion and Effusion. J. Chem. Educ. 44:740–44
11. Hofmann AW (1866) On the Action of the Trichloride of Phosphorus on the Salts of the Aromatic Monoamines. Proc. Roy. Soc. 15:55–62
12. Kolbe H (1871) Ueber die Structurformeln und die Lehre von der Bindung der Atome. J. prakt. Chem. 111:127–36
13. Rocke AJ (2010) Image and Reality: Kekulé, Kopp, and the Scientific Imagination. University of Chicago Press
14. Kolbe H (1870) Ueber einige Abkömmlinge des Cyanamids. J. prakt. Chem. 109:288–306
15. Mach E (1897) Popular Scientific Lectures. Open Court, Chicago
16. Couper AS (1858) On a New Chemical Theory. Phil. Mag. 16:104–16
17. Rocke AJ (1993) The Quiet Revolution: Hermann Kolbe and the Science of Organic Chemistry. Los Angeles: University of California Press
18. Ramberg P (2003) Chemical Structure, Spatial Arrangement: The Early History of Stereochemistry. Ashgate, Aldershot
19. Gmelin L (1843–70) Handbuch der Chemie, 4th ed., 13 vols. Winter, Heidelberg
20. Mitscherlich E (1844) Lehrbuch der Chemie, 4 ed. Mittler, Berlin
21. Rocke AJ (1990) 'Between Two Stools': Kopp, Kolbe and the History of Chemistry. Bull. Hist. Chem. 9:19–24
22. Wislicenus W (1882) Ueber die Schätzung von Haftenenergien der Halogene und des Natriums an organischen Resten. Ann. Chem. 212:239–50
23. Laurent A (1854) Méthode de chimie. Mallet-Bachelier, Paris
24. Russell CA (1971) The History of Valency. Leicester University Press
25. Swain PA (2005) Bernard Courtois (1777–1838). Bull. Hist. Chem. 30:103–11
26. Erlenmeyer E (1871) Die Aufgabe des chemischen Unterrichts. Akademie der Wissenschaften, Munich
27. Rowbottom M, Susskind C (1984) Electricity and Medicine: History of Their Interaction. San Francisco Press, 1984
28. Grotthus T (1806) Sur la décomposition de l'eau et des corps qu'elle tient en dissolution à l'aide de l'électricité galvanique. Ann. chim. 58:54–73
29. Kopp H (1840) Einiges über Löslichkeit. Ann. Chem. 34:260–71
30. Gay-Lussac JL (1839) Considérations sur les forces chimiques. C. r. Acad. Sci. 8:1000–17

Author Biography

Educated first as a chemist, Alan J. Rocke earned a Ph.D. in history of science at the University of Wisconsin-Madison in 1975. Today he is the Henry Eldridge Bourne Professor of History at Case Western Reserve University in Cleveland, Ohio, where he has been teaching since 1978. Rocke is the author of five books and nearly fifty journal articles on the history of chemistry, especially the theoretical development of the science in Europe during the course of the nineteenth century. He is a Fellow of the American Association for the Advancement of Science, and in 2000 received the Dexter Award from the American Chemical Society for lifetime achievement in the history of chemistry.

A. J. Rocke and H. Kopp, *From the Molecular World*, SpringerBriefs in History of Chemistry, DOI: 10.1007/978-3-642-27416-9, © The Author(s) 2012